Mathematical Engineering

Today, the development of high-tech systems is unthinkable without mathematical modeling and analysis of system behavior. As such, many fields in the modern engineering sciences (e.g. control engineering, communications engineering, mechanical engineering, and robotics) call for sophisticated mathematical methods in order to solve the tasks at hand.

The series Mathematical Engineering presents new or heretofore little-known methods to support engineers in finding suitable answers to their questions, presenting those methods in such manner as to make them ideally comprehensible and applicable in practice.

Therefore, the primary focus is—without neglecting mathematical accuracy—on comprehensibility and real-world applicability.

To submit a proposal or request further information, please use the PDF Proposal Form or contact directly: Dr. Thomas Ditzinger (thomas.ditzinger@springer.com)

Indexed by SCOPUS, zbMATH, SCImago.

Dmitrii Legatiuk

Mathematical Modelling by Help of Category Theory

Abstract Perspectives on Models and Coupling of Models

Springer

Dmitrii Legatiuk
University of Erfurt
Erfurt, Germany

ISSN 2192-4732 ISSN 2192-4740 (electronic)
Mathematical Engineering
ISBN 978-3-031-84950-3 ISBN 978-3-031-84951-0 (eBook)
https://doi.org/10.1007/978-3-031-84951-0

This Springer imprint is published by the registered company Springer Nature Switzerland AG
The registered company address is: Gewerbestrasse 11, 6330 Cham, Switzerland

Preface

Mathematical modelling aims to describe physical, chemical, biological, and other phenomena in terms of mathematical expressions and equation. In other words, it seeks to create a *model* that represents the system under consideration. A critical question for the practical use of models is whether the model itself and the results obtained from it can be trusted. A deeper understanding of models and the development of methods for model assessment were central topics of the DFG Research Training Group GRK 1462 "Evaluation of coupled numerical and experimental partial models in structural engineering" [2], where I began my scientific career. It was quickly recognised that most existing model assessment approaches focus primarily on the practical aspects of models, in other words, they address the correctness of model implementation.

Looking more generally at the model creation process, we can observe that models are not immediately provided as mathematical expressions, but rather the result of a formalisation process of specific modelling assumptions. Consequently, the overall model assessment must also address these modelling assumptions and their formalisation. Errors arising from the violation of modelling assumptions are particularly significant because they make the model *conceptually* wrong. Moreover, while the practical use of such a wrong model might still be correct from the algorithmic perspective, the results it produces are meaningless. Consequently, addressing conceptual modelling errors requires a deeper understanding of the model creation process, particularly in the case of coupled problems where multiple models need to be coupled.

The goal of this book is to propose a *structural perspective* on models, the relations between them, and the coupling of models. This perspective necessitates a formal mathematical description of models. Among different possibilities, *category theory* was chosen as the mathematical foundation for this work. For a long time after its introduction by S. Eilenberg and S. Mac Lane in [3], category theory was often referred to by its nickname "abstract nonsense" to emphasise that this theory has no practical applications. However, it is now safe to say that category theory has evolved from a theory with only few applications in algebraic topology to the theory with applications in various fields of science and technology. The last decade was particularly important for expanding the applications areas of category theory. As this book illustrates, category theory provides the flexibility to describe mathematical models in an abstract manner, without delving into specific details of their application domain.

A general context for the rest of this book is established in Chap. 1. This chapter begins with a general discussion on models and modelling process within the engineering context. After that, the model consistency and the modelling errors are discussed. It is also proposed in this chapter to distinguish between the *conceptual modelling errors*, which appear because of violating basic modelling assumptions, and the *instance modelling errors*, which are related to the practical implementation of models. A brief overview of the conceptual modelling approaches is provided at the end of the chapter.

Chapter 2 presents preliminaries from category theory, categorical ontology logs, and the lattice of theories concept. Additionally, an overview of some classical and more recent applications of category theory focusing in particular on applications related to conceptual modelling is also presented in this chapter.

Chapter 3 plays a central role in this book, as it introduces the *category theory-based modelling framework*. In this framework, mathematical models are associated with sets of their underlying modelling assumptions, which are objects of categories of mathematical models. Relations between mathematical models are introduced through the concept of *model complexity*. After discussing individual mathematical models, the framework addresses the coupling of models, demonstrating that the categorical structure for coupled mathematical models is more complicated. This chapter is provided with various illustrative examples for practical interpretations of the abstract constructions introduced in within the category theory-based modelling framework.

The aim of Chap. 4 is to facilitate the practical application of the category theory-based modelling framework by exploring the engineering perspective and providing practical examples. To make the categorical constructions more accessible for practical modelling, their simpler analogues are introduced. Furthermore, model properties, such as for example model uncertainty and model robustness, are discussed and evaluated with respect to their potential for formalisation within the category theory-based modelling framework. This discussion leads to the classification of model properties into *universal model properties* and *applied model properties*. At the end of this chapter, applications of the category theory-based modelling framework to practical problems of bridge aerodynamics and to the development of coupled solution methods are presented.

Chapter 5 offers new directions for extending the ideas developed in the previous chapters. Specifically, it proposes the applications of type theory and abstract algebra to conceptual modelling in engineering. In the beginning of this chapter, ideas for *automatic model generation* are presented, which are based on the formalisation mappings that appear in categories of mathematical models and naturally lead to the use of type theory. The final section of this chapter introduces an alternative perspective on mathematical models combining the lattice of theories concept, categorical ontology logs, and relational algebra approach. This perspective provides more flexibility and can even be applied to define models of physical objects, which are not necessarily described in terms of mathematical expressions and equations.

As typical for new theories and concepts, ideas tend to evolve over time. This is particularly evident when comparing the first publication made on the topic of category theory-based modelling, see [4], with the results presented in this book. Therefore, I believe that the ideas and constructions discussed here will also change and develop further in the future.

When working with formal mathematical theories, it is essential to keep track of the nature of the objects we aim to formally describe. As the motto for the application of category theory to modelling, I would like to quote R. Boute from his very inspiring paper [1]:

> A well-designed axiomatisation is fitted to existing concepts, not the other way around.

I tried to keep this motto in mind during my work, and I hope I managed to follow it for the most of the journey. Of course, this journey would have not been possible without the support of others. I would like to thank especially two persons: Prof. Klaus Gürlebeck and Prof. Henrik Nilsson. Without their support this book would never have been written. Special thanks also go to Prof. Tom Lahmer, who supported me from the beginning of my work at GRK 1462 and carefully reviewed several parts of this book.

Erfurt, Germany Dmitrii Legatiuk

References

1. Boute, R. (2016). Why mathematics needs engineering. *Journal of Logical and Algebraic Methods in Programming, 85*(5), 867–878.
2. DFG Research Training Group GRK 1462. (2008–2017). *Evaluation of coupled numerical and experimental partial models in structural engineering*. Bauhaus-Universität Weimar.
3. Eilenberg, S., & Mac Lane, S. (1945). General theory of natural equivalences. *Transactions of American Mathematical Society, 58*, 231–294.
4. Gürlebeck, K., Hofmann, D., & Legatiuk, D. (2017). Categorical approach to modelling and to coupling of models. *Mathematical Methods in the Applied Sciences, 40*(3), 523–534.

Competing Interests The author has no competing interests to declare that are relevant to the content of this manuscript.

Contents

Chapter 1
Modelling Process in Engineering

Abstract This chapter establishes a general context for the rest of this book. We start with a general discussion on models and modelling process within the engineering context. We then discuss the model consistency and the modelling errors that may arise during the modelling process. Specifically, we emphasise that it is necessary to distinguish between the *conceptual modelling errors*, which appear because of violating basic modelling assumptions, and the *instance modelling errors*, which are related to the practical implementation of models. After that, we provide a brief overview of the conceptual modelling approaches, which could help us to evaluate models with respect to their coherence, as well as provide possibilities for a more formal description of the model creation process.

1.1 Models and Modelling Process

The understanding of models and, especially, mathematical models, is the main objective of this book, and several perspective on models will be provided in the upcoming chapters. Models are nowadays widely used in all areas of science, and there is a general common understanding of what to understand under the term "model". This common understanding has been summarised by R. Aris in the following form [1]:

> In these notes the term "mathematical model" – usually abbreviated to "model" – will be used for any complete and consistent set of mathematical equations which is thought to correspond to some other entity, its prototype. The prototype may be a physical, biological, social, psychological or conceptual entity, perhaps even another mathematical model.

This informal description of a model brings up two points which are crucial for this book:

1. What exactly does a *complete and consistent set of mathematical equations* mean? In other words, is there a way to verify the consistency of mathematical equations in the modelling context?

D. Legatiuk, *Mathematical Modelling by Help of Category Theory*,
Mathematical Engineering, https://doi.org/10.1007/978-3-031-84951-0_1

2. It is explicitly stated, that models are equations, which thought to correspond to some other entity. It is essential to discuss if this perspective on models could be enhanced with more formal definitions and precise formulations.

In fact, this book is entirely devoted to providing a deeper understanding of mathematical models and the modelling process in engineering, addressing these two points by the tools of abstract mathematics and category theory.

Given the growing popularity of machine learning methods in engineering applications, it is necessary to refine the term "model" by distinguishing between two general model types based on their fundamental construction:

- *physics-based models* [15], which are based on clear modelling assumptions that are formalised in terms of mathematical expressions and equations;
- and *data-driven models* [41], which are based on data and represented by e.g. machine learning algorithms, regressions etc., or, in other words, these models are not based on modelling assumptions, but on data.

Evidently, the informal description of a model provided in [1] corresponds to the physics-based models. Throughout this book, we will also associate mathematical models with physics-based models, and we will make always clear if other types of models are considered.

In practice, solving any engineering problem begins not directly with a model, but with a *modelling process* that produces a model for the engineering system under consideration. The ultimate goal of the modelling process, however, is not merely create the model but to analyse it and draw conclusions about the engineering system. Looking more generally at the modelling process, it is safe to assume that the final objective of the modelling process is to describe a certain real-world phenomenon. Hence, independent of the model type, a model is required to describe this real-world phenomenon, or, in other words, approximate the reality [4, 46]. However, before creating a model, it is essential to clearly formulate a modelling task that includes parameters and effects the model must encompass. In this regard, the output of the modelling process—the final model—must be evaluated with respect to the modelling task formulated in the beginning. This model evaluation aims at assessing *model quality*, or, in other words, to "measure" how adequate the model represents the physical phenomenon under consideration. Figure 1.1 schematically illustrates a general framework of the modelling process, indicating that the model quality assessment requires not only the final result (the model itself), but also the initial modelling task; otherwise, criteria for assessing the model would be undefined.

A distinct feature of modern engineering problems is that because of their growing complexity, conventional simplified single-model approaches for engineering problems are increasingly substituted by combinations of coupled models. To provide a structural perspective on these combinations of coupled models, the

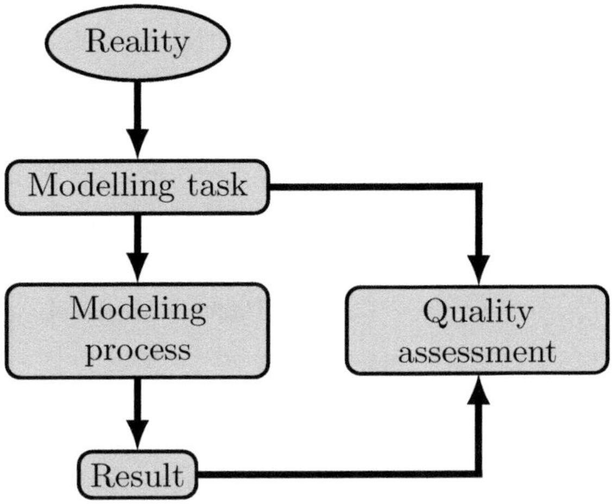

Fig. 1.1 A general framework for model evaluation

Research Training Group 1462, supported by the German Research Foundation (DFG), has propose distinguishing the following model types [11]:

- *partial models* are models which are coupled to other models;
- a *coupled model* is the model coupling several partial models;
- a *global model* is the model of a real object derived from coupled/partial models.

To be precise, if we are interested in modelling of a building, then a single structural element, e.g. a beam, would be a partial model, if we consider several beams constituting one storey, then that would be a coupled model, and if we model a complete building, then we obtain a global model. However, if we are interested in modelling of only a single beam, then the global model would consist of only one partial model in this case. In this context, the primary difference between global model and a coupled model is that a "global model" represents a complete model of a real object, whereas a coupled model may be an incomplete model that combines several partial models. Figure 1.2 presents this general approach of describing a real object with the help of coupled partial models.

The intuitive description of partial and global models provided above can be expressed more formally as follows:

Definition 1.1 A model A is called to be a *partial model* of a model B, if and only if $A \subset B$, but $B \not\subset A$. Moreover, a model M is said to be a *global model*, if and only if there does not exist a model N such that $M \subset N$.

Although this definition, supported by our discussion, is intuitively understandable, it is nonetheless necessary to formally define what is actually meant by a "model" and to explain how different models can be related by the inclusion relation $\subset$, as well as what this relation signifies. Addressing these points is among the main objectives of this books, particularly in the context of coupled models. In this regard, the precise meaning of the inclusion relation $\subset$ in the context of modelling will

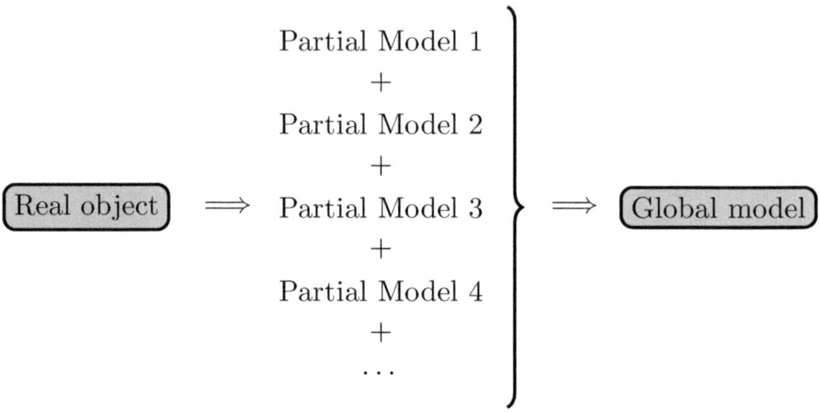

Fig. 1.2 The coupling of different partial models results in a global model of a real object

be discussed in Chap. 3, where *category theory* will be used as the mathematical backbone for a formal description of models and relations between them.

If we examine the modelling process illustrated in Fig. 1.1 more closely, then there are various ideas regarding the steps that constituting this process, see for example [33, 38]. It is also important to emphasise, that the "Modelling process" block in Fig. 1.1 has been singled out for illustrative purposes; however, formulating a modelling task and assessing the result are also integral part of the modelling process. Transferring now the schematic representation of the modelling process depicted in Fig. 1.1 into a sequence of steps, we arrive at the general workflow of the modelling process shown in Fig. 1.3.

Let us provide few remarks on some of the steps shown in Fig. 1.3:

- **Step 2.** The term "physical phenomena" should be understood in a broader context covering problems of a physical, chemical, biological, and other nature, rather than being limited to problems from physics alone.
- **Steps 3–4.** These steps could be combined together if only a single partial model is considered, in which case the global model would be identical with the partial model.
- **Step 5.** The solution methods can be analytical, providing an explicit solution; numerical, offering an approximate solution; or coupled, combining the advantages of both methods.

Additionally, some steps in the general workflow shown in Fig. 1.3 may be optional, depending on choices made in previous steps. For instance, if data-driven models are selected in Steps 3 and 4 only, there is no need to choose specific solution methods, as the approach for performing computations with a data-driven model is, in fact, determined by the model choice itself. Furthermore, each step can be further subdivided into smaller sub-steps, if necessary. Finally, it is safe to say, that the Steps

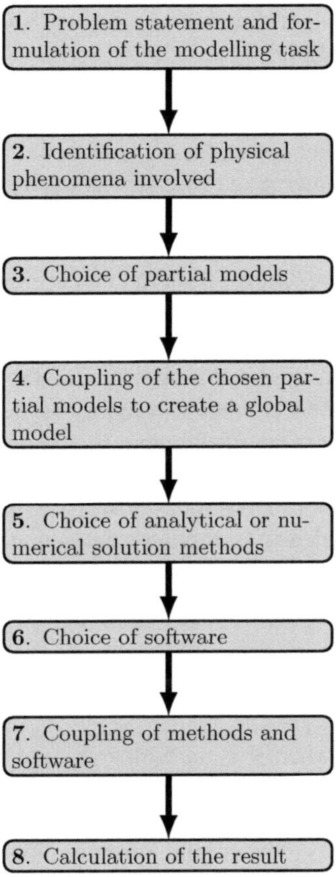

Fig. 1.3 General workflow of the modelling process

1–4 are related to *conceptual modelling* of a problem, and Steps 5–8 are related to a practical use of models, which later will be referred to as *instance modelling*.

1.2 Modelling Errors

Model quality, in the sense discussed above, strongly depends on the formulation of a modelling task. Nonetheless, even if the modelling task is well-formulated, modelling errors can appear on different stages of the modelling process, spanning from the conceptualisation of a model to the interpretation of simulation results. This means that errors affecting the quality of the final result,can occur at each step of the workflow shown in Fig. 1.3. However, the *nature* of these errors can be very different depending on the step at which they arise. For instance, errors arising from misconceptions of physical phenomena appearing in Step 3 are very different from algorithmic errors in Step 5. Consequently, the treatment of these errors must also differ.

When discussing modelling errors, it is also important to remark that, in practical applications, solution methods and the models underlying them are often mixed. For example, finite element models constructed in software packages are often interpreted by engineers as a global model of a real object. In reality, however, these finite element models are simply the result of applying a specific solution method. In this regard, the errors associated with the use of a specific finite element model, generally speaking, can originate from two sources: conceptual modelling errors in Steps 1–4, and errors arising from the practical solution of a problem in Steps 5–8. In that context, it is assumed that the output of modelling Steps 1–4 is a model formulated in terms of mathematical expressions and equations to be solved, and, hence, Steps 5–8 represent the corresponding solution steps.

One of the main motivation for this book is to provide a deeper understanding of conceptual modelling errors appearing during the modelling process, as well as to propose ideas how these errors can be detected. In this regard, it is necessary to clearly formulate what is understood under the term "conceptual modelling error". Therefore, let us consider a trivial illustrative example by studying the classical heat equation given by

$$c\rho\frac{\partial\theta}{\partial t} - \operatorname{div}\left(\lambda(\mathbf{x})\operatorname{grad}\theta\right) = 0, \quad \mathbf{x}\in\mathbb{R}^n,$$

where ϑ is the temperature of a body, ρ is the density of the material, c is the specific heat capacity, and λ is the thermal conductivity. In the case of constant thermal conductivity λ, the partial differential equation presented above can be simplified to its most common formulation by using the Laplace operator

$$c\rho\frac{\partial\theta}{\partial t} - \lambda\Delta\,\theta = 0.$$

Because this formulation is very common, a typical conceptual error, in the case of non-constant thermal conductivity, is to assume that this simplified model formulation is valid also in this case by just making the conductivity a function of $\mathbf{x}$:

$$c\rho\frac{\partial\theta}{\partial t} - \lambda\Delta\,\theta = 0 \qquad \Longrightarrow \qquad c\rho\frac{\partial\theta}{\partial t} - \lambda(\mathbf{x})\Delta\,\theta = 0.$$

The last equation is obviously wrong from the modelling point of view, but in the same time, it makes perfect sense as a partial differential equation that can be solved numerically or analytically.

Another source of conceptual modelling errors is the coupling of different partial models when assembling a global model. Coupling conditions for mathematical models are based on specific modelling assumptions: for example, in the case of modelling the heat conduction process in a domain made of two different materials, the coupling is realised through transmission conditions

$$[u] = \left[\frac{\partial u}{\partial \mathbf{n}} \right] = 0$$

ensuring the continuity of the temperature field and heat flux, $[\cdot]$ denotes the jump of a function. These coupling conditions are related to the Fourier's law of heat propagation within the domain. In this case, a conceptual modelling error would occur if, instead of ensuring the continuity of the normal derivative of the temperature field u, one assumes continuity of the gradient $\operatorname{grad} u$, which might also appear to be a reasonable assumption.

The examples for heat equation indeed illustrate the nature of conceptual errors—basic modelling assumptions are violated at a later stage in the modelling process, or, while working with the models. Additionally, with the growing complexity of engineering problems nowadays, the number of critical points that can lead to conceptual modelling errors, as discussed above, also increases. Because conceptual modelling errors appear on the early stage in the modelling process, their impact on the quality of the final result is much more profound. Moreover, as it has been indicated by the examples, conceptual modelling errors do not necessary make the model formulation wrong (differential equation in the examples), but rather make the results of computations with this model formulation meaningless. Because of that reason, detection of conceptual modelling errors is much more complicated. In particular, it is not possible to look only at the final model formulations for detecting such errors, but rather necessary to understand how the model creation process works and what assumptions have been used, and if they can be alternated without violating the model correctness.

To provide a better understanding of the modelling process, we propose to work with the modelling methodology, distinguishing two broad phases of the modelling process:

- **Conceptual modelling phase**: Concerned with capturing all relevant aspects of the problem under consideration as mathematical formulations.
- **Instance modelling phase**: Concerned with the solution procedure for these formulations, which can be done by means of analytic or numerical methods, or, in general, by means of simulations.

Let us remark, that a similar division of the modelling process in a *formulation phase* and a *solution phase* has been proposed in [46], which are described as follows:

> - In the *formulation phase*, the problem is described using basic principles or governing laws and assumed relations taken from some branches of knowledge, such as physics, biology, chemistry, economics, geometry, probability or others. Then all side-conditions that are needed to completely define the problem must be identified: geometric constrains, initial conditions, material properties, boundary conditions and design parameter values. Finally, the properties of interest, how they are to be measured, relevant variables, coordinate systems and a system of units must all be decided on.

- Then, in the *solution phase*, mathematical modelling provides approaches to reformulating the original problem into a more convenient structure from which it can be reduced into solvable parts that can ultimately be re-assembled to address the main questions of interest for the problem.

The essential difference between the formulation phase and the conceptual modelling phase is that, in the conceptual modelling phase, the outcome is a model formulated in terms of mathematical expressions and equations without reference to concrete units, measurements, or quantities of interest, as is done in the formulation phase. These latter points are then addressed in the instance modelling phase. The reason for this fine tuning is motivated by the fact that the same differential equation can describe different physical phenomena depending on the specific practical application. In this regard, the subdivision of the modelling process into two phases can be understood as follows: the conceptual modelling phase deals with abstract objects called "models", which are then *instantiated* for specific practical applications in the instance modelling phase. Chapters 3–5 are devoted to discussing these ideas in detail.

Taking into account the subdivision of the modelling process into two phases, the modelling errors can be then classified accordingly into:

- **Conceptual errors**: Related to the coherence of the model as a whole;
- **Instance errors**: Concrete errors related to practical aspects of modelling; e.g. questions of approximation quality, implementation, scaling problems, and correct geometric modelling.

From this classification, we can conclude that even if the highest possible quality is achieved in the instance modelling phase, the overall model quality may still be affected by errors coming from the conceptual modelling: for example if some basic physical assumptions were forgotten during the coupling of partial models. Especially, without any way of catching conceptual errors, the fact that there is an error would in this case not become manifest until it is discovered that the model simply does not approximate the reality sufficiently well during model validation. Moreover, the validation does not provide clear hints on where the problem actually lies, resulting in time-consuming checks of the complete model. This situation is perfectly described by the following sentence:

It is perfectly possible to formulate conceptually **wrong** model, but *solve it correctly* by numerical methods.

The work presented in this book aims to establish a foundation for the modelling concept, enabling identification of conceptual modelling errors. However, as it has been mentioned above, conceptual modelling errors detection first requires an understanding of the model creation process. Given that because of the complexity of modern engineering problems, coupled models are more and more popular nowadays, it is of utmost importance to understand how coupling of models works at the conceptual level. Therefore, the main part of this work is devoted to development of a conceptual modelling framework based on category theory, which provides a clear structure for relations between different models and their couplings. Nonetheless, some first ideas towards conceptual modelling error detection, as well as automatic model generation, will also be presented.

1.3 Evaluation of Models and Conceptual Modelling in Engineering

When working with models that describe several physical phenomena, it is essential to understand a general *structure* of the models and their coupling. Without this understanding, it is impossible to assess whether the final model is free from conceptual modelling errors or to evaluate the model's quality with respect to the fulfilment of the original modelling task. Consequently, it is not surprising that in recent years, several attempts to develop modelling frameworks that clearly describe the *structure* of coupled models, along with methodologies for evaluation and guidance of engineering modelling, have been performed. In this section, without pretending to be complete, we provide a short overview (not in a chronological order) of existing approaches to conceptual modelling and model evaluation in engineering. Moreover, we review these approaches with respect to their applicability to the conceptual modelling phase and their potential for detecting conceptual modelling errors, or providing hints towards this detection.

1.3.1 Engineering Approaches to Model Evaluation

A general framework for identifying error and uncertainties in computational simulations has been proposed in [33]. In this framework, the modelling process of a physical system has been subdivided into the following six phases:

1. conceptual modelling of the physical system;
2. mathematical modelling of the conceptual model;
3. discretisation and algorithm selection for the mathematical model;
4. computer programming of the discrete model;
5. numerical solution of the computer program model;
6. representation of the numerical solution.

Comparing these phases with the two-phase modelling methodology proposed in the previous section, it becomes clear that phases 1 and 2 belong to the conceptual modelling phase, and phases 3–6 belong to the instance modelling phase. Because the focus of this book is to propose new perspective on conceptual modelling through the tools of abstract mathematics and category theory, let us recall the key aspects of phases 1 and 2 from [33]:

> During the *conceptual modelling phase*, no mathematical equations are written, but the fundamental assumptions regarding possible events and physical processes are made. Only conceptual issues are considered, with heavy emphasis placed on determining all possible factors, such as physical and human intervention, that could possibly affect the requirements set for the modelling and simulation.
>
> The primary activity in the *mathematical modelling of the conceptual model* phase is to develop detailed and precise mathematical models. Any complex mathematical model of a process, or physical system is actually composed of many mathematical submodels. The complexity of the models depends on the physical complexity of each phenomenon being considered, the number of physical phenomena considered, and the level of coupling of different types of physics.

Although this description of the two phases is intuitively clear, a formal definition of models, and particularly of the term "model complexity" has not been introduced. As we will see throughout this book, adopting a broader perspective on the conceptual modelling phase supports the introduction of formal definitions of mathematical models and their complexity, because the models cannot be considered separately from their initial conceptions. In the latter case, we face the situation that the same boundary value problem for a differential equation (a model in the sense of [33]) may be used to describe different physical phenomena depending on the context, i.e. depending on the model conceptualisation. This knowledge will be formalised by the help of category theory in Chap. 3.

Because the main objective of the work [33] was to develop a general framework for identifying error and uncertainty in computational simulations, general sources of uncertainties have been described on each of six phases. In particular, two types of uncertainties have been addressed:

- *aleatory uncertainty*—uncertainty in model parameters,
- and *epistemic uncertainty*—inaccuracy due to lack of knowledge.

In a broader context, the epistemic uncertainty could be seen as a kind of general name for conceptual modelling errors. However, it is rather difficult to formally define and specify the model "inaccuracy due to lack of knowledge", and, therefore, we will keep using the term "conceptual modelling errors" in this book. Moreover, considering that the overall goal of work [33] was to support engineering modelling in practice, it is natural that theoretical question related to the conceptual modelling could not be discussed in full generality. As a result, the question of coupling and comparing partial models during conceptual modelling, which is critical for assessing overall model quality, has not been addressed.

Another important remark is that in practice, conceptual modelling is often treated as the initial step in the modelling process, as discussed above, and once, for example, a differential equation is formulated, the conceptual modelling phase is generally "forgotten". Consequently, if verification and validation procedures indicate a necessity for model updates, these updates are typically made in later stages of the modelling process, although all model updates must be consistent with the original conceptualisation of the model.

The modelling framework, proposed in [33], has been further developed in [31, 32]. In particular, the ideas for quality assessment of models based on the concept of *verification and validation* have been introduced. According to [3], these terms are described as follows

- *verification* is associated with the quality of numerical treatments of mathematical models, and
- *validation* addresses the quality of mathematical models.

However, neither the consistency of the modelling process, nor the coupling quality are clearly addressed by verification and validation. The reason is that the verification and validation process checks only the correctness of the global model, but cannot find the reasons why the model is not satisfactory.

Looking at engineering approaches to quality assessment of models, it is necessary to mention the approach proposed in [17], where the so-called *predictive maturity* of models has been introduced. A definition of a model, however, is very different to the ones discussed above:

> In this paper, the terms "model", "code", and "simulation" are used inter-changeably to denote an analytical or numerical model. The are collectively referred to as the "model" that, in short, builds a functional relation between inputs and outputs:
>
> $$y = M(p; \vartheta),$$
>
> where the pair $(p; \vartheta)$ denotes inputs to the model and y is the output.

From this definition, it is evident that the conceptual modelling phase is not addressed by the predictive maturity concept. In general, this concept was developed to assess how closely the model prediction aligns with experimental test results. In some sense, the predictive maturity concept can be seen as a complement to general validation and verification procedures. Despite of this concept being extremely useful for practical engineering, it is not applicable to address theoretical aspects of the modelling process arising in the conceptual modelling phase.

An approach to estimating modelling errors in computational mechanics has been proposed in [34], utilising tools of functional analysis. However, in this and related works, the modelling error has been treated as the discretisation error in quantity of interest and has been estimated by help *a posteriori* error estimates. Thus, the concept proposed in [34] does not address conceptual errors appearing in the modelling process, but rather focuses on the quality of numerical treatment of a problem. Examples and counterexamples of working with *a posteriori* error estimates for assessing quality of models were presented in [28].

1.3.2 Graph Theory-Inspired Evaluation of Models and Their Couplings

From analysis of the methodology proposed in [31–33] and of the concept of verification and validation, it became evident that the overall quality assessment of models requires a more *structural* perspective on models, as well as an evaluation of a coupling between partial models as well. A first attempt to provide a structural perspective on engineering models from the field of civil engineering has been proposed within the framework of the DFG Research Training Group GRK 1462 "Evaluation of coupled numerical and experimental partial models in structural engineering" [11] by utilising graph theory to represent the construction of a global model.

Inspired by the definition of a directed weighted graph and various applications of graph theory, see for example [19], graph theory has been applied to describe connections between different models by a graph in works [18, 21, 22]. In these works, a concept for model quality evaluation has been proposed, where partial models are represented by vertices of a directed graph and the coupling relations between the partial models are provided by the edges. Figure 1.4 illustrates this idea on the example of three different model classes.

In the framework proposed in [18, 21, 22], global models are obtained by identifying paths in a weighted graph along the vertices and edges. Hence, the weights of vertices b_i represent the quality of partial models, while the arc weights c_{kl} represent the coupling quality. Moreover, the coupling quality was assumed to have the following representation

$$c_{kl} := \begin{cases} 1 & \text{for a perfect coupling between partial models,} \\ \in (0, 1) & \text{coupling with loss of quality,} \\ 0 & \text{if partial models are not coupled.} \end{cases}$$

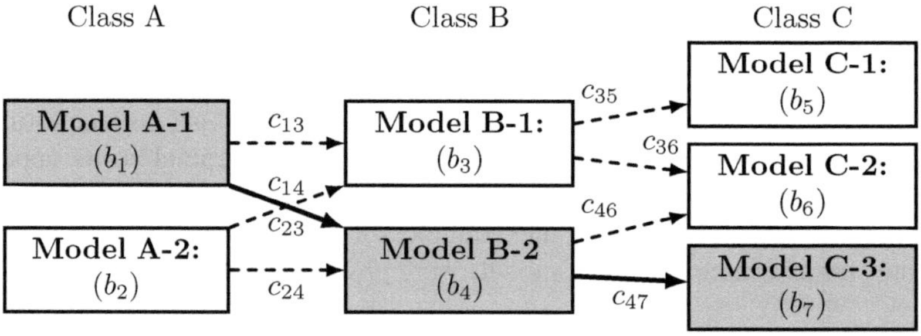

Fig. 1.4 Application of graph-theoretic approach to the modelling, adapted from [21]. Weights b_i represent quality of partial models, and arc weights c_{kl} represent quality of the coupling

Consequently, the global model's quality is then defined as follows

$$MQ_{GM} := \frac{1}{2} \sum_{k \in IN} \sum_{l \in IN} c_{kl} b_l + \sum_{k \in ON} \sum_{l \in ON} c_{kl} b_l, \tag{1.1}$$

where IN denotes the inner nodes, and ON denotes the outer nodes.

To evaluate the model quality in the framework of graph-theoretic approach, model sensitivity and model uncertainty have been used as measures. Moreover, model uncertainty has been understood as parameter uncertainty (e.g. uncertainty of structural properties or of material constants), as it has been proposed in [25], which is not really related to the classical uncertainty principle appearing in theoretical physics and time-frequency analysis, e.g. [16]. Precisely, the model quality b_l is calculated by the help of sensitivity indices according to the formula

$$b_l = \frac{S_{Ti,l}^{MC} \cdot MQ_{PM_l}}{\sum_{i=1}^{n_{N,red}} S_{Ti}^{MC}} \quad \text{with } PM_l \in M_i,$$

where M_i is the model class, $n_{M,red} = \left|\left\{M_i \mid S_{Ti}^{M_i} \geq 0.3\right\}\right|$ is the number of classes of partial modes with a non negligible influence on the global response, $\sum_{i=1}^{n_{N,red}} S_{Ti}^{MC}$ is a normalising constant, and $S_{Ti,l}^{MC}$ is the total sensitivity index, which is generally defined as follows

$$S_{Ti} := 1 - \frac{V(E(Y|X_i))}{V(Y)},$$

where $V(E(Y|X_i))$ is the variance of the expected value of Y when conditioning with respect to all parameters except for X_i.

Based on the ideas presented above, it can be summarised that the global sensitivity analysis is used to quantify the influence of partial models on the output of the global model. Similar ideas are applied to individual classes of partial models, which are then ranked with respect to their quality. Additionally, it is essential to evaluate also the coupling quality. In work [21], the coupling quality between different partial models has been assumed to be perfect, i.e. $c_{kl} = 1$ for all edges in a graph. In work [22], the coupling quality has been addressed in more details:

> Within the scope of this paper, coupling is defined as data coupling and the quality of coupling is related to the quality of data transfer. Let α be a model quantity appearing in both partial model k and l at the same point on the structure, for example the support force of a column, which is present in the column itself and also at the top of the foundation. A perfect data coupling ensures consistent data in both models, $\alpha^k = \alpha^l$, which refers to data coupling quality of $cq_{\alpha,k-l} = 1$. As the differences in transferred data increases, which means that $\alpha^k \neq \alpha^l$, the quality of the coupling decreases down to a quality of zero when

no data is transferred, $\alpha^k \neq 0$ and $\alpha^l = 0$. This leads to the following definition of data coupling quality:

$$cq_{\alpha,k-l} = 1 - \frac{\left|\alpha^k - \alpha^l\right|}{\max\left\{\left|\alpha^k\right|, \left|\alpha^l\right|\right\}}.$$

The definition of coupling proposed in [22] is motivated by a computer science perspective, where coupling quality is typically understood as the quality of data transfer between different software systems, see [14] and references therein. Additionally, the application of graph theory restricts the coupling quality to be represented by a single number, as discussed above. In [22], an idea of calculating this number based on model sensitivity and uncertainty has been proposed. However, it is difficult to interpret a practical meaning of this number for physics-based models: for instance, in the case of modelling the heat conduction process in a domain made of two different materials, the coupling is realised via transmission conditions

$$[u] = \left[\frac{\partial u}{\partial \mathbf{n}}\right] = 0$$

ensuring continuity of the temperature field and heat flux. In this situation, it is rather impossible to assign a meaningful number to this coupling. Nonetheless, in the context of data-driven models, the strategy of assigning a number to the coupling quality is makes perfect sense, as data-driven models are typically coupled in the sense of data transfer: the output of one model serves as the input for another.

It is also necessary to mention that the model sensitivity proposed in [18, 21, 22] as a model quality measure is straightforward to compute and interpret in practical engineering applications. However, it lacks *universality* in the sense, that the model sensitivity depends on specific boundary and initial conditions. It is easy to construct situations where, for the same model but with different boundary/initial conditions, the significance of various model parameters differs. In this regard, the graph theory-based framework for model assessment coupled with the model sensitivity addresses engineering practice, but is not "abstract enough" to address also the conceptual modelling of engineering systems or to provide hints towards conceptual modelling errors detection.

Looking more generally at the graph-theoretic approach, it can be noticed that the graph theory has been mostly used as a graphical representation of models and *connections* between them, where weights of this graphs were referred to as quality, see again Fig. 1.4. The quality of global model is then calculated according to (1.1) as a path through the graph, however the strategy of working with discrete optimisation problems, typical for applications of graph theory [19], for finding the highest possible quality of the global model has not been adapted. Working with discrete optimisation problems would work perfectly well when the graph-theoretic approach is used to evaluate quality of data-driven models, as it has been mentioned above. Nonetheless, the idea of using diagrams (graphs) to represent models and connections between them is fundamental to this book. The difficulty of using the graph-theoretic approach for the conceptual modelling comes from the limitations of graph theory: it lacks more advanced mathematical structures to adequately describe models and their couplings. A proper mathematical formalism for an abstract description of models would be the category theory [2, 23], as it will be shown in Chap. 3. The core idea is that all numerical quantities assigned to edges and vertices are waived, and the meaning of edges is addressed by more abstract constructions relating objects in categories.

1.3.3 Models and Abstract Hilbert Spaces

A general perspective on models based on functional analysis, which represent systems in different fields of science, has been presented [12, 13]. In particular, it has been shown that if specific conditions are satisfied, then there is a Hilbert space corresponding to such a system. The conditions for a model of a system are the following:

1. The system is represented by a fixed number N of variables $(X_k)_{k=1}^N$;
2. Each variable belongs to an open subset O_k of a separable Fréchet real vector space V_k;
3. At least one of the vector spaces $(V_k)_{k=1}^N$ is infinite dimensional;
4. For any other model of the system using N variables $\left(X'_k\right)_{k=1}^N$ belonging to open subset O'_k of V_k, and for $X_k, X'_k \in O_k \cap O'_k$ there is a continuous map $X'_k = F_k(X_k)$.

Additionally, it is assumed that the variables are independent in the sense that there is no relation such as $X_k = R(X_l)$. Moreover, the derivate $\dfrac{dX_k}{dt}$ of a variable X_k is considered as an independent variable. All variables must belong to vector spaces and can be restricted to take only some range, and qualitative variables represented by discrete values are excluded at this stage. Nonetheless, the theory presented in [12, 13] addresses also such variables with more refined constructions. The vector spaces are infinite dimensional whenever the variables are functions: for example when they represent the time-evolution of the system—in this case, X_k is the function itself $X_k : \mathbb{R} \to O_k :: X_k(t)$.

After introducing the conditions for a model of a system, the following proposition has been proved [12]:

Proposition 1.1 *For any system represented by a model meeting the conditions 1–4, there is a separable, infinite dimensional, Hilbert space H, defined up to isomorphism, such that S can be embedded as an open subset $\Omega \subset H$ which contains* 0 *and a convex subset.*

After that the ideas of a basis and the complexification of Hilbert spaces have been introduced.

Finally, let us recall also another important notion from [12]—*observables*:

> The idea here is to replace X by another variable $\Phi(X)$, that is called an observable, which still belongs to the same vector space V, so there is a map $\Phi\colon V \to V$. If X is the actual value of the variables (and of the state), then $\Phi(X)$ is the function which is estimated by the statistical procedure: the data are used to compute the value of the parameters of $\Phi(X)$. We make three genera assumptions about Φ:
>
> **Definition 1.2** Definition of an observable:
>
> 1. an observable is a linear map: $\Phi \in L(V; V)$;
> 2. the range of an observable is a finite dimensional vector subspace of V: $\dim \Phi(X) < \infty$;
> 3. $\forall X \in O$, $\Phi(X)$ is an admissible value, that is $\Phi(O) \subset O$.

From this discussion, it is visible that the main motivation for proposing a conceptual modelling framework based on Hilbert spaces comes from the theoretical physics—particularly particle physics—where observables and Hilbert spaces are commonly used. However, applying this framework to describe engineering problems requires further study and analysis. Especially the practical interpretation of abstract variables and Hilbert spaces could be problematic for engineers in real-world applications. Nonetheless, the idea of using abstract mathematics to develop a systematic view on the conceptual modelling discussed in [12, 13] is central for this book, and was a starting point to look for a suitable mathematical formalism.

1.3.4 Abstract Approaches to System Modelling

Speaking about other mathematical formalisms, that could serve as a backbone for a formal description of models, relations between them, and model evaluations, it is natural to also consider model theory [36] and logic, which represent the *meta-mathematics*. However, the application of model theory and the lower predicate calculus typically requires extensive formal constructions, making it less applicable in practice, particularly for models described by partial differential equations. Nonetheless, in the field of modelling of dynamical systems, predicate logic and metalanguages have shown their usefulness, see [43–45] and references therein. In these works, it has been shown that system models can be turned into logical equations, resulting in the fact that many important properties of these models can be proved by studying the corresponding logical equations. In this book, we will not use model theory and predicate logic to address the conceptual modelling phase. Instead, we will work with similar formal theories, such as type theory [35] and lambda calculus [5], in Chap. 5, where these theories are utilised to detect conceptual modelling errors in practice. Furthermore, considering that there is a connection between logic and type theory, then it is possible to see that the type theory-based approach is a kind of alternative to [43–45], and that perhaps there is a way to utilise advantages of both by connecting them.

When discussing conceptual modelling of engineering systems, it is necessary to mention another class of approaches that use abstract algebraic definitions of individual system parts and combine them by the help of relations to describe a complete system, see for example works [27, 39, 42]. Chapter 5 extents these ideas to a *meta-metamodelling* level, enabling the conceptual modelling of complex engineering systems at a highly abstract level, focusing on the structure of the system, rather on technical details.

Finally, it is important to note that the use of category theory for system modelling has garnered significant attention in recent years, see e.g. [8, 9]. Because using category theory to conceptual modelling in engineering applications is a central theme of this book, an overview of classical and modern applications of category theory will be provided in Chap. 2, where all necessary mathematical preliminaries will be introduced.

1.3.5 Modelling of Engineering Systems by Means of Modelling Languages

Modelling languages are especially popular in system engineering, because since the 1980s, it is widely known that engineering systems can be developed more easily by concentrating on the semantics of the application domain using *conceptual modelling languages* [7]. Conceptual modelling languages allow capturing the semantics of applications through formal notations in order to achieve a more natural and direct modelling of aspects of the engineering systems. Common conceptual modelling languages include, for example, the Entity-Relationship model, the Enhanced Entity-Relationship model, Unified Modeling Language class diagrams, Object-Role Modeling (ORM), and Second Generation ORM [20]. However, by their construction, conceptual modelling languages do not address the issue of the modelling errors detection, because the modelling process is not really addressed by the conceptual modelling languages. Additionally, the applicability of modelling languages to describe a general semantic of physics-based models is not entirely clear.

Practice-oriented modelling languages, such as Simulink [40] or Modelica [24], support the work of engineers by combining conceptual capabilities and practical modelling. Modern modelling languages actually do support modelling at a high level of abstraction, through (undirected) equations or equivalent diagrams, but they are still primarily intended for practical modelling: ultimately the goal is to generate working simulation (or other) code. Consequently, many aspects of such languages are geared towards supporting that goal, meaning that models often become less abstract and more operational than they could be. Moreover, the kinds of equations supported are limited, such as differential-algebraic equations in the case of Modelica with some limited extensions to partial differential equations [37], which are numerically treated. That means that many problems in the wider setting of mathematical physics, like boundary value problems of elasticity theory, fluid mechanics, or electrodynamics, cannot be modelled adequately as partial differential equations and integral equations are necessary in these cases.

To enable conceptual modelling errors detection by means of modelling languages, it is necessary to *formalise* them via introduction stricter semantics and rules. Several attempts in this direction have been presented in recent years: see for example works [26] and [6] for applications in the fields of cyber-system modelling and Unified Modelling Language, respectively. In the context of this book, the use of type theory is particularly important. In the context of modelling languages it means, that the type system of a modelling language must be appropriately designed: by static checking of model derivation, it can be ensured that the conceptual modelling has been carried out such that the resulting multiphysics model does not contradict basic physical laws and assumptions before preparing simulation code. Originating from the Functional Reactive Programming [29], Functional Hybrid Modelling (FHM) [10, 30] is a first step in the development of a Haskell-based modelling language with a robust type system allowing us to detect modelling errors more effectively compared to the type systems in the traditional modelling languages. In essence, FHM enriches Haskell with a notion of first-class equations for modelling of physical systems. Haskell provides the required abstraction mechanisms and means for constructing systems by composing equation system fragments. While, in terms of language design and modelling aims, FHM is similar to Modelica and thus suffers the same limitations, the design of FHM can serve as a blueprint for a language for modelling systems in the abstract, demonstrating the usefulness of Haskell for this kind of modelling language research and design. First steps in extension of the FHM to realistic models appearing in engineering will be presented in Chap. 5.

1.4 Summary of the Chapter and Objectives of the Book

The solution of any engineering problem starts with the modelling process that aims to adequately model a given physical phenomenon. Considering that the conceptual modelling of a problem is the first step in this process, conceptual modelling errors can have a severe impact on the entire modelling process. Detecting conceptual modelling process requires a more formal approach to models and relations between them. This is particularly crucial when considering coupled models, because coupling itself can be a source of potential conceptual modelling errors.

In recent years, various approaches have been proposed to develop conceptual modelling frameworks that provide a deeper understanding of models and their couplings have been made in recent years. However, despite the many advantages provided by these approaches, a unified approach formally describing physics-based models, relations between them, and coupling between them has yet not been presented. This book presents an abstract conceptual modelling framework based on category theory to address this need.

This book aims at addressing the following objectives:

1. Develop a more *structural* point of view on mathematical models by help of category theory, where by "structural point of view" we understand a formal mathematical structure describing mathematical models.
2. Discuss how different models can be related to one another.
3. Provide a formal basis for describing the coupling of mathematical models in the language of category theory.
4. Analyse if some abstract categorical constructions can be made meaningful in the modelling context.
5. Enhance engineering applications of the category theory-based modelling framework by providing clear practical meaning for abstract definitions, as well as discussing model comparison in practice.
6. Present first ideas towards automatic detection of conceptual modelling errors.
7. Propose a meta-metamodelling approach based on abstract algebra and category theory to address system modelling in engineering applications.

References

1. Aris, R. (1994). *Mathematical modelling techniques*. Dover Publications Inc.
2. Awodey, S. (2010). *Category theory*. Oxford University Press Inc.
3. Babuška, I., & Oden, J. T. (2004). Verification and validation in computational engineering and science: Basis concepts. *Computer Methods in Applied Mechanics and Engineering, 193*, 4057–4066.
4. Babuška, I., & Oden, J. T. (2005). The reliability of computer predictions: Can they be trusted? *International Journal of Numerical Analysis and Modeling, 1*, 1–18.
5. Barendregt, H. P. (1984). *The lambda calculus. Its syntax and semantics*. Elsevier Science Publishers B.V.
6. Borges, R. M., & Mota, A. C. (2007). Integrating UML and formal methods. *Electronic Notes in Theoretical Computer Science, 184*, 91–112.
7. Borgida, A. (1986). Conceptual modeling of information systems. In M. L. Brodie, J. Mylopoulos, & J. W. Schmidt (Eds.), *Topics in information systems — On knowledge base management systems* (pp. 461–469). Springer.
8. Breiner, S., Subrahmanian, E., & Jones, A. (2017). Categorical foundations for system engineering. In A. M. Madni, B. Boehm, D. A. Erwin, R. Ghanem, & M. J. Wheaton (Eds.), *Proceedings of 15th Annual Conference on Systems Engineering Research*.
9. Breiner, S., Sriram, R. D., & Subrahmanian, E. (2019). Compositional models for complex systems. In *Artificial intelligence for the Internet of everything* (pp. 241–270). Elsevier.
10. Capper, J. (2014). *Semantic methods for functional hybrid modelling*. PhD thesis, University of Nottingham.
11. DFG Research Training Group GRK 1462. (2008–2017). Evaluation of coupled numerical and experimental partial models in structural engineering. Bauhaus-Universität Weimar.
12. Dutailly, J. C. (2014). *Hilbert spaces in modelling of systems* (47 pp.). HAL Id: hal-00974251.
13. Dutailly, J. C. (2014). *Common structures in scientific theories* (34 pp.). HAL Id: hal-01003869.
14. Fröbel, T., Firmenich, B., & Koch, C. (2011). Quality assessment of coupled civil engineering applications. *Advanced Engineering Informatics, 25*, 625–639.
15. Gershenfeld, N. (1998). *The nature of mathematical modeling*. Cambridge University Press.

16. Gröchenig, K. (2000). *Foundations of time-frequency analysis*. Birkhäuser.
17. Hemez, F., Atamturktur, H. S., & Unal, C. (2010). Defining predictive maturity for validated numerical simulations. *Computers and Structures, 88*, 497–505.
18. Jung, B. (2015). *Evaluation of coupled partial models for the assessment of restrain effects in concrete structures*. PhD thesis, Bauhaus-Universität Weimar.
19. Jungnickel, D. (2008). *Graphs, networks and algorithms*. Springer.
20. Keet, C. M. (2008). A formal comparison of conceptual data modeling languages. In *Proceedings of the 13th International Workshop on Exploring Methods in Systems Analysis and Design*, Montpellier, June 16.
21. Keitel, H., Karaki, G., Lahmer, T., Nikulla, S., & Zabel, V. (2011). Evaluation of coupled partial models in structural engineering using graph theory and sensitivity analysis. *Engineering Structures, 33*, 3726–3736.
22. Keitel, H., Jung, B., Motra, H. B., & Stutz, H. (2014). Quality assessment of coupled partial models considering soil-structure coupling. *Engineering Structures, 59*, 565–573.
23. Mac Lane, S. (1978). *Categories for the working mathematician*. Springer.
24. Modelica. (2012). *Modelica – A unified object-oriented language for systems modelling; Language specification version 3.3*. Modelica Association.
25. Möller, D., & Beer, M. (2004). *Fuzzy randomness. Uncertainty in civil engineering and computational mechanics*. Springer.
26. Neema, S., Simko, G., Levendovszky, T., Porter, J., Agrawal, A., & Sztipanovits, J. (2014). Formalization of software models for cyber-physical systems. In *ACM Proceedings of the 2nd FME Workshop on Formal Methods in Software Engineering* (pp. 45–51).
27. Nefzi, B., Schott, R., Song, Y. Q., Staples, G. S., & Tsiontsiou, E. (2015). An operator calculus approach for multi-constrained routing in wireless sensor networks. In *Proceedings of the 16th ACM International Symposium on Mobile Ad Hoc Networking and Computing*, New York, June 22–25.
28. Nikulla, S. (2012). *Quality assessment of kinematical models by means of global and goal-oriented error estimation techniques*. PhD Thesis, Bauhaus-Universität Weimar.
29. Nilsson, H., Curtney, A., & Peterson, J. (2002). Functional reactive programming, continued. In *Proceedings of the 2002 ACM SIGPLAN Haskell Workshop* (pp. 51–64).
30. Nilsson, H., Peterson, J., & Hudak, P. (2003). Functional hybrid modeling. In *Lecture notes in computer science* (Vol. 2562, pp. 376–390). Springer.
31. Oberkampf, W. L., & Roy, C. J. (2010). *Verification and validation in scientific computing*. Cambridge University Press.
32. Oberkampf, W. L., & Trucano, T. G. (2002). Verification and validation in computational fluid dynamics. *Progress in Aerospace Sciences, 38*(3), 209–272.
33. Oberkampf, W. L., DeLang, S. M., Rutherford, B. M., Diegert, K. V., & Alvin, K. F. (2002). Error and uncertainty in modeling and simulation. *Reliability Engineering and System Safety, 75*, 333–357.
34. Oden, J. T., & Prudhomme, S. (2002). Estimation of modeling error in computational mechanics. *Journal of Computational Physics, 182*, 496–515.
35. Pierce, B. C. (2002). *Types and programming languages*. MIT Press.
36. Robinson, A. (1963). *Introduction to model theory and to the metamathematics of algebra*. North-Holland Publishing Company.
37. Saldamli, L., & Fritzon, P. (2000). Object-oriented modeling with partial differential equations. In *Proceedings of Modelica Workshop 2000* (pp. 119–128).
38. Samarskii, A. A., & Michailov, A. P. (2005). *Mathematical modelling: Ideas. Methods. Examples*. Phismalit (in Russian).

39. Simko, G., Levendovszky, T., Maroti, M., & Sztipanovits, J. (2014). Towards a theory for cyber-physical systems modeling. In *Proceedings of the 4th ACM SIGBED International Workshop on Design, Modeling, and Evaluation of Cyber-Physical Systems*, Berlin, April 14.
40. SIMULINK. (1992). *SIMULINK – User's guide*. MathWorks, Inc.
41. Solomatine, D., See, L. M., & Abrahat, R. J. (2008). Data-driven modelling: Concepts, approaches and experiences. In *Practical hydroinformatics, water science and technology library*. Springer.
42. Syarif, A., Abouaissa, A., Idoumghar, L., Lorenz, P., Schott, R., & Staples, S. G. (2019). New path centrality based on operator calculus approach for wireless sensor network deployment. *IEEE Transactions on Emerging Topics in Computing, 7*(1), 162–173.
43. Vassilyev, S. N. (2006). Method of reduction and qualitative analysis of dynamic systems: I. *Journal of Computer and Systems International, 45*(1), 17–25.
44. Vassilyev, S. N., Davydov, A. V., & Zherlov, A. K. (2008). Intelligent control via new efficient logics. In *Proceedings of the 17th World Congress The International Federation of Automatic Control*, Seoul, July 6–11.
45. Vassilyev, S. N., Druzhinin, A. E., & Morozov, N. Yu. (2015). Derivation of preservation conditions for properties of mathematical models. *Doklady Mathematics, 92*(3), 658–663.
46. Witelski, T., & Bowen, M. (2015). *Methods of mathematical modelling. Continuous systems and differential equations*. Springer International Publishing.

Chapter 2
Preliminaries from Category Theory

Abstract This chapter presents mathematical concepts from category theory, which constitute the basis for the results discussed in the upcoming chapters. We start with the introduction of basic ideas of category theory and related definitions. After that, constructions on categories and abstract structures are defined. Next, *lattice of theories* concept is also important for some parts of this book, therefore it will be introduced in this chapter as well. After that, this chapter presents categorical ontology logs, or simply *ologs*, which combine flexibility of classical all-purpose ontologies and mathematical rigorosity of category theory. Finally, we discuss some classical and more recent applications of category theory focusing in particular on applications related to conceptual modelling.

2.1 Basics of Category Theory

Category theory was introduced in 1945 by S. Eilenberg and S. Mac Lane in [14] as a mathematical *concept* devoting to provide a common language for describing different mathematical structures (objects) and relations between them. In general words, category theory provides a language which helps to describe properties of different objects by using their representation with diagrams of arrows. In this case, each arrow represents a function (mapping) between two objects. Before starting with basic definitions from category theory, we would like to recall two important principles of category theory [25]:

1. Objects should not be defined or characterised by intrinsic properties, but rather by their relations with other objects. Objects that are isomorphic in terms of relations therefore cannot be distinguished.
2. Constructions can be reflexively iterated, i.e., when constructions are performed with objects at a certain level, we can pass to a higher level, where the objects are constructions of the previous level. Put differently, the relations between objects at one level can become the objects at the next level.

These principles illustrate the biggest advantage and, in the same time, disadvantage of category theory: although the objects need still to be clearly defined, their internal

D. Legatiuk, *Mathematical Modelling by Help of Category Theory*,
Mathematical Engineering, https://doi.org/10.1007/978-3-031-84951-0_2

structure is effectively "forgotten" and the relations (mappings) between the objects essentially defines the structure of category. In this regard, category theory provides tools to develop a structural view on different fields by "sieving out" all non-essential details. But if after establishing this structural view a detailed look at the objects is required, then more refined tools need to be used. We could view this situation in some sense similar to the conceptual and instance modelling phases discussed in Chap. 1: category theory provides means to address the conceptual modelling phase, and the instance modelling phase requires are tools and methods.

Next we will present some basic definitions from category theory. The material presented in this and further sections is essentially based on the classical and modern works in category theory [1, 3, 29, 38], all details on the constructions presented here can be found in these works. We start our discussion with a formal definition of a category:

Definition 2.1 A *category* consists of the following data:

- *Objects*: $A, B, C, \ldots$
- *Arrows (morphisms)*: $f, g, h, \ldots$
- For each arrow f, there are given objects $\mathrm{dom}(f)$ and $\mathrm{cod}(f)$ called the *domain* and *codomain* of f, respectively. We write

  $$f \colon A \longrightarrow B \quad \text{or} \quad A \xrightarrow{f} B$$

  to indicate that $A = \mathrm{dom}(f)$ and $B = \mathrm{cod}(f)$.
- Given arrows $f \colon A \longrightarrow B$ and $g \colon B \longrightarrow C$, that is, with $\mathrm{cod}(f) = \mathrm{dom}(g)$, there is given an arrow

  $$g \circ f \colon A \longrightarrow C$$

  called the *composite* of f and g.
- For each object A, there is given an arrow

  $$1_A \colon A \longrightarrow A$$

  called the *identity arrow* of A.

These data are required to satisfy the following laws:

- Associativity: $h \circ (g \circ f) = (h \circ g) \circ f$ for all $f \colon A \longrightarrow B$, $g \colon B \longrightarrow C$, $h \colon C \longrightarrow D$.
- Unit: $f \circ 1_A = f = 1_B \circ f$ for all $f \colon A \longrightarrow B$.

A category is anything that satisfies this definition; and moreover, in general, the objects do not have to be sets and the arrows need not be functions. In this sense, a category is an abstract algebra of functions, or arrows (sometimes also called *morphisms*), with the composition operation $\circ$ as primitive.

Before introducing other construction on categories and abstract structures, let us briefly point out few examples of so-called *concrete categories*, which are informally categories in which the objects are sets, possibly equipped with some structure, and the arrows are certain, possible structure-preserving, functions, see again [3, 29] for the details:

- **Set**: objects are all small sets and arrows are all functions between them.
- **Set**$_{\text{fin}}$: objects are all finite sets and arrows are all functions between them.
- **Grp**: objects are all small groups and arrows are all group homomorphisms;
- **Top**: objects are topological spaces and arrows are continuous mappings between them.

A special role in the applications of category theory discussed in this book play *finite categories*. Classical examples of finite categories are the following:

- **0** is the empty category with no objects and no arrows;
- **1** is the category with one object and one (identity) arrow;
- **2** is the category with two objects a, b and just one arrow $a \to b$ not the identity;
- **3** is the category with three objects whose non-identity arrows are arranged as in the triangle

In addition to the categories mentioned above, abstract categories in which arrows are not "functions" play even more prominent role in applications, because of a higher flexibility in constructing categories. Among all different categories, we would like to explicitly point out the following category:

Definition 2.2 The category **Rel** has sets as objects and has binary relations as arrows. That is, an arrow $f: A \longrightarrow B$ is an arbitrary subset $f \subseteq A \times B$. The identity arrow on a set A is the identity relation,

$$1_A = \{(a, a) \in A \times A \mid a \in A\} \subseteq A \times A.$$

Given $R \subseteq A \times B$ and $S \subseteq B \times C$, define composition $S \circ R$ by

$$(a, c) \in S \circ R \text{ iff } \exists\, b\colon (a, b) \in R \wedge (b, c) \in S$$

that is, the "relative produce" of S and R.

The category **Rel** will play a special role in the construction of the category theory-based modelling framework presented in this book. Strictly speaking, application of category theory to mathematical modelling is generally based on a specific construction of objects and arrows in **Rel**. Nonetheless, for the clarity of application context, we will introduce separate names for our categories, and we will illustrate a precise construction of these categories with practical examples from applied mathematical modelling.

Next, we define another special type of categories—*discrete categories*:

Definition 2.3 A category is *discrete* when every arrow is an identity.

Let X be a set, and if we add one identity arrow $x \rightarrow x$ for each $x \in X$, then X is the set of objects of a discrete category, and every discrete category is so determined by its set of objects. Thus, discrete categories are sets.

Equivalent objects in a category are defined in the categorical sense as follows:

Definition 2.4 In any category **C**, an arrow $f: A \rightarrow B$ is called an *isomorphism*, if there is an arrow $g: B \rightarrow A$ in **C** such that

$$g \circ f = 1_A \text{ and } f \circ g = 1_B.$$

We say that A is *isomorphic* to B, if there exists an isomorphism between them.

Next step is to introduce relations between different categories, i.e. the arrows or morphisms between categories. A mapping between categories is called a functor:

Definition 2.5 A *functor*

$$F : \mathbf{C} \longrightarrow \mathbf{D}$$

between categories **C** and **D** is a mapping of objects to objects and arrows to arrows, in such a way that

1. $F(f: A \longrightarrow B) = F(f): F(A) \longrightarrow F(B)$,
2. $F(1_A) = 1_{F(A)}$,
3. $F(g \circ f) = F(g) \circ F(f)$.

That is, F respects domains and codomains, identity arrows, and composition.

Functors can be further specified as follows:

- A functor which simply "forgets" some or all of the structure of an object is commonly called a *forgetful functor* (or, an *underlying functor*). For example, the functor U: **Grp** $\rightarrow$ **Set**, assigning to each group G the set UG of its element and to each morphism $f: G \rightarrow G'$ of groups the same function f, regarded just as a function between sets, is forgetful, because the group structure is omitted.
- A functor T: **C** $\longrightarrow$ **B** is *full* when to every pair C, C' of objects of **C** and to every arrow $g: TC \longrightarrow TC'$ of **B**, there is an arrow $f: C \longrightarrow C'$ of **C** with $g = Tf$.
- A functor T: **C** $\longrightarrow$ **B** is *faithful* (or an embedding) when to every pair C, C' of objects of **C** and to every pair $f_1, f_2: C \longrightarrow C'$ of parallel arrows of **C** the equality $Tf_1 = Tf_2: TC \longrightarrow TC'$ implies $f_1 = f_2$.

A morphism of functors is defined by a natural transformation:

Definition 2.6 For categories $\mathbf{C}$, $\mathbf{D}$ and functors $F, G\colon \mathbf{C} \longrightarrow \mathbf{D}$ a *natural transformation* $\vartheta\colon F \longrightarrow G$ is a family of arrows in $\mathbf{D}$

$$(\vartheta_C\colon FC \longrightarrow GC)_{C \in \mathbf{C}},$$

such that, for any $f\colon C \longrightarrow C'$ in $\mathbf{C}$, one has $\vartheta_{C'} \circ F(f) = G(f) \circ \vartheta_C$, that is, the following diagram commutes:

$$\begin{array}{ccc} FC & \xrightarrow{\vartheta_C} & GC \\ {\scriptstyle Ff}\downarrow & & \downarrow{\scriptstyle Gf} \\ FC' & \xrightarrow[\vartheta_{C'}]{} & GC' \end{array}$$

Given such a natural transformation $\vartheta\colon F \longrightarrow G$, the $\mathbf{D}$-arrow $\vartheta_C\colon FC \longrightarrow GC$ is called the *component of* ϑ at C.

2.2 Constructions on Categories and Abstract Structures

Category theory originates from the definitions of basic objects and structures introduced in the previous section. Next step is to introduce new constructions on categories by using the basic objects. Thus, new categories are produced from the "old" allowing a more detailed study of categories. Some of abstract structures and constructions on categories are introduced in this section.

We start with the definition of products of categories:

Definition 2.7 The *product* of two categories $\mathbf{C}$ and $\mathbf{D}$, written as

$$\mathbf{C} \times \mathbf{D}$$

has objects of the form (C, D), for $C \in \mathbf{C}$ and $D \in \mathbf{D}$, and arrows of the form

$$(f, g)\colon (C, D) \longrightarrow (C', D')$$

for $f\colon C \longrightarrow C' \in \mathbf{C}$ and $g\colon D \longrightarrow D' \in \mathbf{D}$. Composition and units are defined componentwise, that is

$$(f', g') \circ (f, g) = (f' \circ f, g' \circ g)$$
$$1_{(C,D)} = (1_C, 1_D).$$

There are two obvious *projection functors*

$$\mathbf{C} \xleftarrow{\pi_1} \mathbf{C} \times \mathbf{D} \xrightarrow{\pi_2} \mathbf{D}$$

defined by $\pi_1(C, D) = C$ and $\pi_1(f, g) = f$, and similarly for π_2.

Functors and their transformations naturally give rise to a special category:

Definition 2.8 Given the categories **B** and **C**, then a category which has as objects the functors $\mathbf{T}\colon \mathbf{C} \to \mathbf{B}$, and as morphisms the natural transformations between two of such functors is called a functor category $\mathbf{B}^{\mathbf{C}} = \mathrm{Funct}(\mathbf{C}, \mathbf{B})$.

The next step is to introduce an abstract characterisation of the empty set and the one-element sets in the category **Sets** (objects are sets, infinite in general, and arrows are all functions between them) and structurally similar objects in general categories. This characterisation is done via *initial* and *terminal* objects:

Definition 2.9 In any category **C**, and object

- 0 is *initial* if for any object C there is a unique morphism $0 \longrightarrow C$,
- 1 is *terminal* if for any object C there is a unique morphism $C \longrightarrow 1$.

Initial and terminal objects are uniquely determined up to isomorphism [3]:

Proposition 2.1 *Initial (terminal) objects are unique up to isomorphism.*

Next, we need to introduce a notion of a *subcategory*:

Definition 2.10 A *subcategory* **S** of a category **C** is a collection of some of the objects and some of the arrows of **C**, which includes with each arrow f both the object $\mathrm{dom} f$ and the object $\mathrm{cod} f$, with each object s its identity arrow $1_{\mathbf{S}}$ and with each pair of composable arrows $s \longrightarrow s' \longrightarrow s''$ their composite.

The conditions, listed in the definition of a subcategory, ensure that the collections of objects and arrows themselves constitute a category **S**. Moreover, the injection (inclusion) map $\mathbf{S} \longrightarrow \mathbf{C}$ which sends each object and each arrow of **S** to itself (in **C**) is a functor, the *inclusion functor*. This inclusion functor is automatically faithful. We say that **S** is a *full subcategory* of **C** when the inclusion functor $\mathbf{S} \longrightarrow \mathbf{C}$ is full.

Further, we recall the definition of a *comma category*. However, due to a quite technical construction of a general case, we provide a complete derivation of it following [29]. If b is an object of the category **C**, the category of *objects under* b is the category $(b \downarrow \mathbf{C})$ with objects all pairs $\langle f, c\rangle$, where c is an object of **C** and $f\colon b \longrightarrow c$ an arrow of **C**, and with arrows $h\colon \langle f, c\rangle \longrightarrow \langle f', c'\rangle$ those arrows $h\colon c \longrightarrow c'$ of **C** for which $h \circ f = f'$. Thus an object of $(b \downarrow \mathbf{C})$ is just an arrow in **C** from b and an arrow of $(b \downarrow \mathbf{C})$ is a commutative triangle with top vertex b. In a diagrammatic form:

objects $\langle f, c\rangle$: 

arrows $\langle f, c\rangle \xrightarrow{h} \langle f', c'\rangle$: 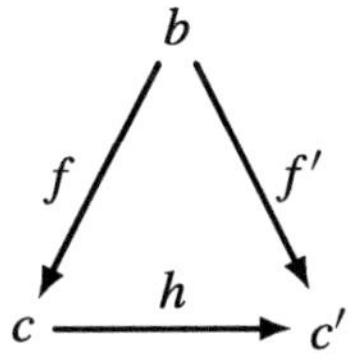

The composition of arrows in $(b \downarrow \mathbf{C})$ is then given by the composition in $\mathbf{C}$ of these base arrows h of these triangles.

If a is an object of $\mathbf{C}$, the category $(\mathbf{C} \downarrow a)$ of *objects over* a has

objects: 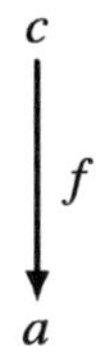

arrows: 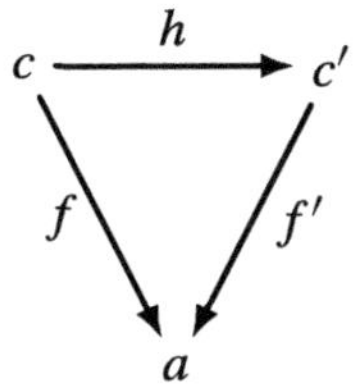

the triangle commutative.

If b is an object of $\mathbf{C}$ and $S : \mathbf{D} \longrightarrow \mathbf{C}$ a functor, the category $(b \downarrow S)$ of *objects S-under b* has as objects all pairs $\langle f, d\rangle$ with d being an object of $\mathbf{D}$ and $f : b \longrightarrow Sd$ and as arrows $h : \langle f, d\rangle \longrightarrow \langle f', d'\rangle$ all those arrows $h : d \longrightarrow d'$ in $\mathbf{D}$ for which $f' = Sh \circ f$. In diagrams,

objects: 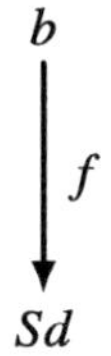

arrows h: 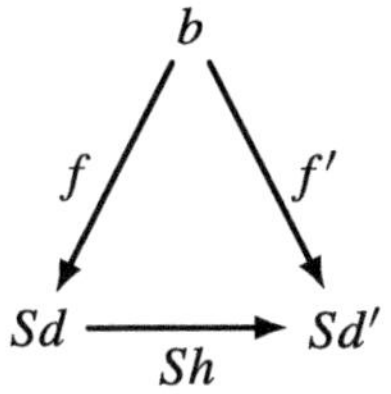

Composition is given by composition of the arrows h in $\mathbf{D}$. Again, if $a \in \mathbf{C}$ and $T : \mathbf{E} \longrightarrow \mathbf{C}$ is a functor, one may construct a category $(T \downarrow a)$ of *objects T-over* a.

Now we are ready to formulate a general construction of a comma category:

Definition 2.11 Given categories and functors

$$\mathbf{E} \xrightarrow{T} \mathbf{C} \xleftarrow{S} \mathbf{D}$$

the *comma category* $(T \downarrow S)$, also written (T, S), has as objects all triples $\langle e, d, f\rangle$, with d being an object of $\mathbf{D}$, e being an object of $\mathbf{E}$, and $f : Te \longrightarrow Sd$, and as arrows $\langle e, d, f\rangle \longrightarrow \langle e', d', f'\rangle$ all pairs $\langle k, h\rangle$ of arrows $k : e \longrightarrow e'$, $h : d \longrightarrow d'$ such that $f' \circ Tk = Sh \circ f$. In diagrams,

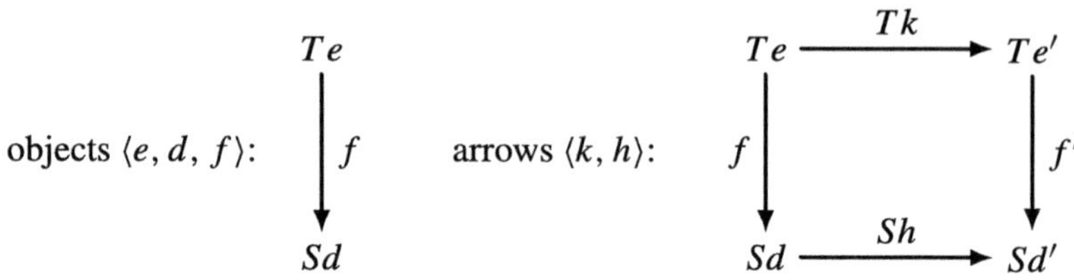

with the square commutative. The composite $\langle k', h'\rangle \circ \langle k, h\rangle$ is $\langle k' \circ k, h' \circ h\rangle$, when defined.

Next we recall the definition of a universal arrow:

Definition 2.12 If $S\colon \mathbf{D} \to \mathbf{C}$ is a functor and c an of object of C, a *universal arrow* from c to S is a pair $\langle r, u\rangle$ consisting of an object r of $\mathbf{D}$ and an arrow $u\colon c \to Sr$ of $\mathbf{C}$, such that to every pair $\langle d, f\rangle$ with d an object of $\mathbf{D}$ and $f\colon c \to Sd$ and arrow of $\mathbf{C}$, there is a unique arrow $f'\colon r \to d$ of $\mathbf{D}$ with $Sf' \circ u = f$. In other words, every arrow f to S factors uniquely through the universal arrow u, as in the commutative diagram

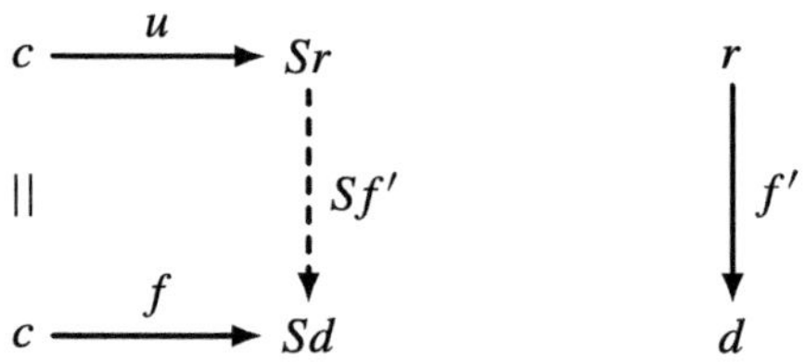

Equivalently, $u\colon c \to Sr$ is universal from c to S when the pair $\langle r, u\rangle$ is an initial object in the comma category $(c \downarrow S)$, whose objects are the arrows $c \to Sd$. As with any initial object, it follows that $\langle r, u\rangle$ is unique up to isomorphism in $(c \downarrow S)$; in particular, the object r of $\mathbf{D}$ is unique up to isomorphism in $\mathbf{D}$.

One of the distinctive feature of category theory is the *duality principle*, which is, simply speaking, the process of reversing all arrows in a category. Formally, the duality principle is given by the following two propositions, see again [3, 29]:

Proposition 2.2 (Formal Duality) *For any sentence Σ in the language of category theory (CT), if Σ follows from the axioms for categories, then so does its dual Σ^*:*

$$CT \Longrightarrow \Sigma \text{ implies } CT \Longrightarrow \Sigma^*.$$

Proposition 2.3 (Conceptual Duality) *For any statement Σ about categories, if Σ holds for all categories, then so does the dual statement $\Sigma*$.*

In Chap. 3, we will use in particular the dual structure of product—the *coproduct*, which is introduced as follows [29]:

Definition 2.13 For any category $\mathbf{C}$, the diagonal function $\Delta\colon \mathbf{C} \to \mathbf{C} \times \mathbf{C}$ is defined on objects by $\Delta(c) = \langle c, c\rangle$, on arrows by $\Delta(f) = \langle f, f\rangle$. A universal arrow from an object $\langle a, b\rangle$ of $\mathbf{C} \times \mathbf{C}$ to the functor Δ is a called a *coproduct diagram*. It consists of an object c of $\mathbf{C}$ and an arrow $\langle a, b\rangle \to \langle c, c\rangle$ of $\mathbf{C} \times \mathbf{C}$; that is, a pair of arrows $i : a \to c$, $j : b \to c$ from a and b to a common codomain c. When such a coproduct diagram exists, the object c is necessarily unique (up to isomorphism in $\mathbf{C}$); it is written $c = a + b$ and is called a *coproduct object*. The coproduct diagram then s

$$a \xrightarrow{i} a + b \xleftarrow{j} b;$$

the arrows i and j are called the *injections* of the coproduct $a + b$ (though they are not required to be injective as functions).

Finally, let us recall the following definition of an equaliser:

Definition 2.14 In any category $\mathbf{C}$, given parallel arrows

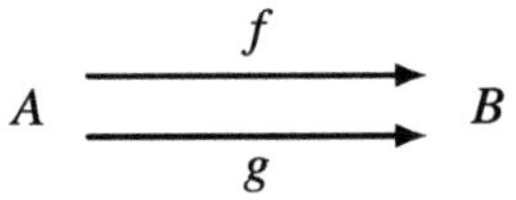

an *equaliser* of f and g consists of an object E and an arrow $e \colon E \to A$, universal such that

$$f \circ e = g \circ e.$$

That is, given any $z \colon Z \to A$ with $f \circ z = g \circ z$, there is a *unique* $u \colon Z \to E$ with $e \circ u = z$, all as in the diagram

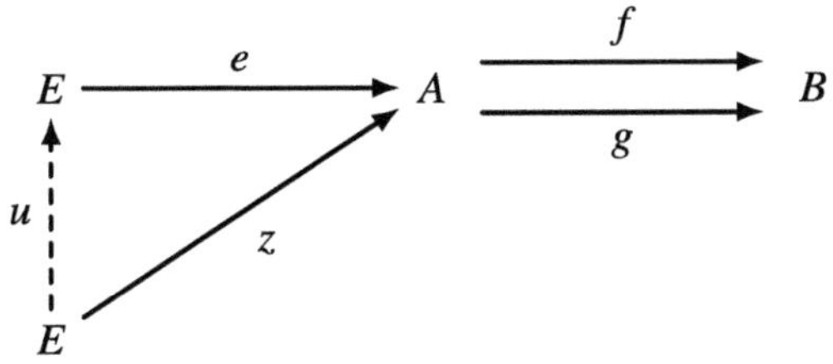

2.3 Lattice of Theories

In Chaps. 3 and 5, the *lattice of theories* concept presented in [35] will be used. The general idea of the lattice of theories comes from the fact that there exists infinitely many theories that can be used to describe any subject and to "organise" these theories, i.e. to see how they are related and how one can come from one theory to another, the lattice of theories has been introduced.

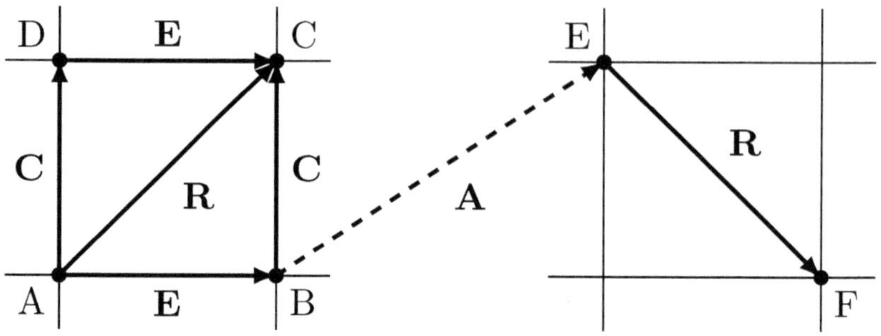

Fig. 2.1 Lattice of theories with four operations to move along the lattice: contraction **C**, expansion **E**, revision **R**, and analogy **A** (adapted from [35])

From each theory, the partial ordering of the lattice defines paths to more general theories above and more specialised theories below. The movements along the paths from one theory to another is realised via *contraction*, *expansion*, *revision*, and *analogy*. Let us briefly recall definitions of these operations:

- *Contraction*. Any theory can be contracted or reduced to a smaller, simpler theory ba deleting one or more axioms.
- *Expansion*. Any theory can be expanded by adding one or more axioms to it.
- *Revision*. At first, some axioms are discarded in a contraction step, and then new axioms are added in an expansion step. The order of steps can be interchanged, if necessary.
- *Analogy*. Unlike contraction and expansion, which move to nearby theories in the lattice, analogy jumps to a remove theory. Analogy then consists in a systematic renaming the types, relations, and individuals that appear in the axioms.

Figure 2.1 illustrates the lattice of theories concept together with the four operations.

By repeated contraction, expansion, and analogy, any theory or collection of beliefs can be converted into any other. Multiple contractions would reduce theory to the empty or universal theory at the top of the lattice that contains only the tautologies that are true of everything. Multiple expansions would lead to the inconsistent absurd theory at the bottom of the lattice, which contains all axioms and is true of nothing.

2.4 Categorical Ontology Logs

In this section, we briefly recall the basic idea of categorical *ontology logs*, or simply *ologs*, which have been proposed by D. Spivak and R. Kent in [37]. The main idea of ologs is to combine the flexibility of classical all-purpose ontologies with the mathematical rigorousity of category theory. The main advantages of using ologs can be summarised as follows:

- ologs are convertible into a database scheme, meaning that the information represented by an olog can easily be input into a computer;
- ologs can be extended, contracted, or revised, if some information must be added or removed, hence, ologs support the lattice of theories concept;
- tools of category theory provide means how to operate with ologs, for example functors can be used to compare different ologs.

We refer to works [36, 37] for a detailed information on ologs. We summarise basic ideas on constructing ologs in the form of the following definition:

Definition 2.15 An olog is a category, which has types as objects, aspects as arrows, and facts as commutative diagrams. The types, aspects, and facts are defined as follows:

- A *type* is an abstract concept represented as a box containing a *singular indefinite noun phrase*. Types are allowed to have compound structure, i.e., being composed of smaller units. Each of the following boxes is a type:

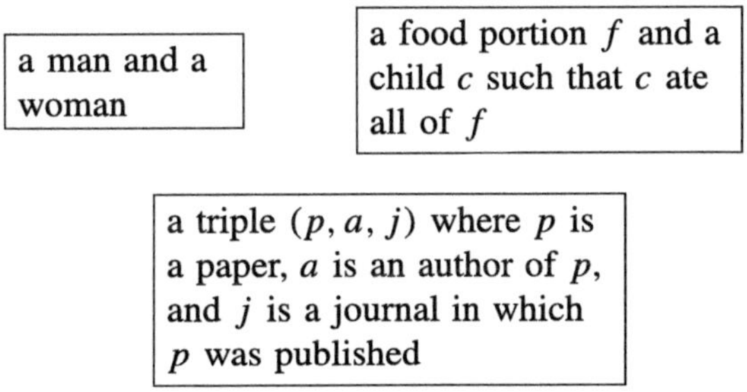

- *Aspects* are functional relationships between the types represented by labelled arrows in ologs. Consider a functional relationship called f between X and Y, which can be denoted $f : X \to Y$, then X is called the *domain of definition* for the aspect f, and Y is called the *set of result values* of f. Here are two examples of using aspects:

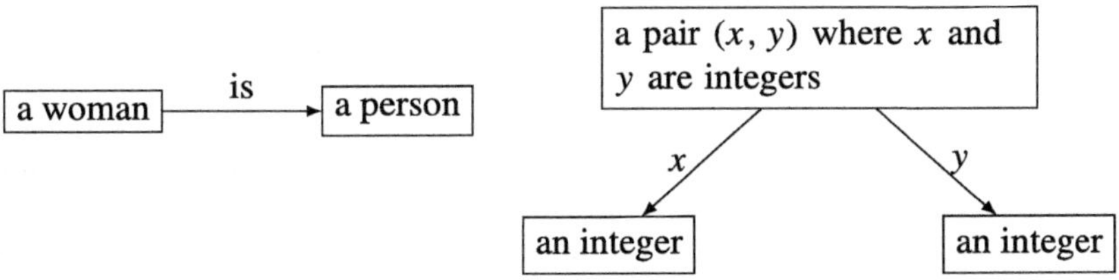

- *Facts* are commutative diagrams, i.e., graphs with some declared path equivalences, in ologs. Facts are constructed by composing several aspects and types. Here is an example of a fact:

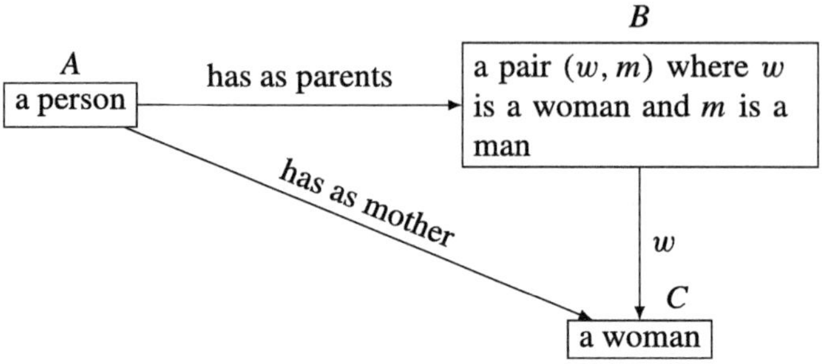

Ologs will be used in Chap. 5 to introduce a *meta-metamodelling* approach for conceptual modelling of engineering system, which are not immediately described by physics-based models. Moreover, to enhance objectively of the approach, ologs will be combined with abstract algebraic definitions.

2.5 Applications of Category Theory

We conclude this chapter with a short overview of classical and modern applications of category theory. As we have already pointed out, category theory originated from the work of S. Eilenberg and S. Mac Lane "General theory of natural equivalences" [14]. Initially, the main areas of applications of category theory were algebraic topology and abstract algebra. However, over time, the general nature of category theory has been appreciated not only in mathematics but also in various other fields of sciences, such as logic, computer science, philosophy, linguistic, system engineering, and many other areas.

In this section, we present a short overview of applications of category theory. However, the aim is not to name each specific contribution or example, but rather to indicate the richness of category theory as a tool for conceptual modelling. Nonetheless, in some cases, we will delve into certain applications in greater detail, especially if they align with the overall scope of the book.

As any conceptual theory, category theory, after its introduction, has been studied with respect to its applicability to formalise mathematics. Thus, one of the first applications of category theory apart from the algebraic topology was foundation of mathematics. For example, in [20] the notion of a *topos* has been introduced. The idea was to view the abstract concept of an arrow in a category as a replacement of the set membership relation. With this idea in mind, the logic has been analysed from the categorical point of view. Additionally, it is worth mentioning that there exists a *category of proofs* for a deductive system of logic [3]. The objects in this category are formulae:

$$\varphi, \psi, \ldots$$

An arrow from φ to ψ is a deduction of ψ from the (uncancelled) assumption φ:

$$\frac{\varphi}{\vdots}$$
$$\overline{\psi}$$

with composition of arrows given by putting together such deductions.

Considering foundational studies of category theory, an approach to the classical functional analysis from categorical perspective has been proposed in [23]. In this work, category theory has been used as an alternative to the classical set theory, and thus the classical functional analysis has been developed from this point of view. Nowadays, category theory is seen in a more general context than "only" foundations for mathematics, but rather as a possible foundation for science in general, see works [27, 30] and reference therein for the details. Additionally, in [22], category theory has been used for development of a framework for concept discovery in science and technology. This work underlines the fundamental nature of category theory as a possible basis for science.

Another important area of applications of category theory is theoretical computer science [5, 7, 34, 39]. In this area of applications, category theory is used as a formal theory to describe programming languages and, as the result, to reason about programs. Particularly, functional programming paradigm essentially reflects ideas of category theory, and we will discuss it a bit more in Chap. 5. Because some results presented in this book were inspired by the connection of category theory and functional programming, let us now recall some more concrete examples from [3] of using category theory in this context:

- Given a functional programming language L, there is an associated category, where the objects are the data types of L, and the arrows are computable functions L, i.e. programs. The composition of two such programs $X \xrightarrow{f} Y \xrightarrow{g} Z$ is given by applying g to the output of f. The identity is the "do nothing" program.
- Type theory is a backbone of the ideas for conceptual modelling error detection presented in Chap. 5. Type theory is mathematically formalised by the help of the *simply typed λ-calculus*, see also [4, 32], which consists of
  - Types: $A \times B, A \to B, \ldots$ (generated from some basic types);
  - Terms:

$$\begin{aligned}
x, y, z, \ldots &: A \text{ (variables for each type } A)\\
a\colon A, b\colon B, \ldots &\text{ (possibly some typed constants)}\\
\langle a, b\rangle &: A \times B \ (a\colon A, b\colon B)\\
\mathrm{fst}(c) &: A \ (c\colon A \times B)\\
\mathrm{snd}(c) &: B \ (c\colon A \times B)\\
ca &: B \ (c\colon A \to B, a\colon A)\\
\lambda x.b &: A \to B \ (x\colon A, b\colon B)
\end{aligned}$$

– Equations:

$$\begin{aligned} \mathrm{fst}(\langle a, b\rangle) &= a \\ \mathrm{snd}(\langle a, b\rangle) &= b \\ \langle \mathrm{fst}(c), \mathrm{snd}(c)\rangle &= c \\ (\lambda x.b)a &= b[a/x] \\ \lambda x.cx &= c \text{ (no } x \text{ in } c) \end{aligned}$$

The relation $a \sim b$ (called *$\beta\eta$-equivalence*) on terms is defined to be the equivalence relation generated by the equations, and renaming of bound variables:

$$\lambda x.b = \lambda y.b[y/x] \text{ (no } y \text{ in } b)$$

The category of types $\mathbf{C}(\lambda)$ is now defined as follows:

– Objects: the types,
– Arrows $A \to B$: closed terms $c\colon A \to B$, identified if $c \sim c'$;
– Identities: $1_A = \lambda x.x$, where $x\colon A$;
– Composition: $c \circ b = \lambda x.c(bx)$.

Additionally, it is worth to mention that there is a connection between the simply typed λ-calculus and the propositional logic known as the *Curry-Howard isomorphism*. Practically it means, that there is a way to formally translate the results from one formalism to another. From the category theory perspective, this the Curry-Howard isomorphism means that there is a functor from the category of proofs in the propositional calculus into the category of types of the λ-calculus, see [3] for details.

Possibilities of using category theory for modelling have been studied by several authors. Especially, applications of category theory to model-driven engineering have been proposed in [13, 26, 28]. However, the most attention in these works has been drawn to software engineering from a computer science point of view. Several authors have also studied the use of category theory to applications in system engineering, see for example works [2, 9, 10, 15, 31]. These works clearly show the usefulness of category theory when describing complex systems. Moreover, it is interesting to note that the approaches of using category theory for modelling purposes vary depending on the authors' ideas. This fact underlines the fruitfulness and flexibility of category theory even when considering the same area of applications.

Categorical ontology logs have been already discussed in this chapter. Nonetheless, we would like to mention here some practical applications of ologs: using ologs in material design has been illustrated in [17], and the link to computer science by connecting ologs with databases has been further studied in [40]. Furthermore, using ologs in a combination with abstract relational algebra for conceptual modelling of engineering systems will be presented in Chap. 5.

Finally, in the end of this section, we would like to mention few other applications of category theory without detailed comments around them and without pretending to be complete:

- probability theory [12, 18, 24] and analysis [21];
- modelling of dynamical systems [6, 19, 33]—the general concept originates from classical algebraic studies of dynamical systems;
- applications in machine leaning [8, 11, 16].

In summary, category theory has been utilised in various applications: from algebraic topology and foundations of mathematics till ontological representation of knowledge and database design. This broad spectrum of applications clearly indicates that category theory is not considered any more as "abstract nonsense" (this term was very often used as a synonym for "category theory"), but became an important field providing tools to develop structural perspectives in various practical applications.

References

1. Adámek, J., Herrlich, H., & Strecker, G. E. (1990). *Abstract and concrete categories. The joy of cats*. Wiley and Sons, Inc.
2. Aguado, E., Gómez, V., Hernando, M., Rossi, C., & Sanz, R. (2024). Category theory for autonomous robots: The marathon 2 use case. In L. Marques, C. Santos, J. L. Lima, D. Tardioli, & M. Ferre (Eds.), *Robot 2023: Sixth Iberian Robotics Conference. ROBOT 2023*. Lecture notes in networks and systems (Vol. 976, pp. 39–52). Springer.
3. Awodey, S. (2010). *Category theory*. Oxford University Press Inc.
4. Barendregt, H. P. (1984). *The lambda calculus. Its syntax and semantics*. Elsevier Science Publishers B.V.
5. Barr, M., & Wells, C. (1990). *Category theory for computing science*. Prentice Hall.
6. Behrisch, M., Kerkhoff, S., Pöschel, R., Schneider, F. M., & Siegmund, S. (2017). Dynamical systems in categories. *Applied Categorical Structures, 15*(1), 29–57.
7. Bird, R., & de Moor, O. (1997). *Algebra of programming*. Prentice Hall.
8. Bradley, T.-D., Terilla, J., & Vlassopoulos, Y. (2022). An enriched category theory of language: From syntax to semantics. *La Matematica, 1*, 551–580.
9. Breiner, S., Subrahmanian, E., & Jones, A. (2017). Categorical foundations for system engineering. In A. M. Madni, B. Boehm, D. A. Erwin, R. Ghanem, & M. J. Wheaton (Eds.), *Proceedings of 15th Annual Conference on Systems Engineering Research*.
10. Breiner, S., Sriram, R. D., & Subrahmanian, E. (2019). Compositional models for complex systems. In *Artificial intelligence for the Internet of everything* (pp. 241–270). Elsevier.
11. Culbertson, J., & Sturtz, K. (2013). *Bayesian machine learning via category theory*. arXiv:1312.1445.
12. Culbertson, J., & Sturtz, K. (2014). A categorical foundation for Bayesian probability. *Applied Categorical Structures, 22*(4), 647–662.

13. Diskin, Z., & Maibaum, T. (2012). Category theory and model-driven engineering: From formal semantics to design patterns and beyond. In *Proceedings of ACCAT 2012 EPTCS 93* (pp. 1–2).
14. Eilenberg, S., & Mac Lane, S. (1945). General theory of natural equivalences. *Transactions of American Mathematical Society, 58*, 231–294.
15. Foley, J. D., Breiner, S., Subrahmanian, E., & Dusel, J. M. (2021). Operands for complex system design specification, analysis and synthesis. *Proceedings of the Royal Society A, 477*, 20210099.
16. Fong, B., Spivak, D. I., & Tuyéras, R. (2021). Backprop as functor: A compositional perspective on supervised learning. In *LICS 19: Proceedings of the 34th Annual ACM/IEEE Symposium on Logic in Computer Science*.
17. Giesa, T., Spivak, D. I., & Buehler, M. J. (2012). Category theory based solution for the building block replacement problem in materials design. *Advanced Engineering Materials, 14*(9), 810–817.
18. Giry, M. (1982). A categorical approach to probability theory. In *Categorical aspects of topology and analysis*. Lecture notes in mathematics (Vol. 915). Springer.
19. Giunti, M., & Mazzola, C. (2012). Dynamical systems on monoids: Towards a general theory of deterministic systems and motion. In G. Minati, M. Abram, & E. Pessa (Eds.), *Methods, models, simulations and approaches towards a general theory of change* (pp. 173–185).
20. Goldblatt, R. (1984). *Topoi. The categorical analysis of logic*. Elsevier Science Publishers.
21. González-Cervantes, J. O., Luna-Elizarrarás, M. E., & Shapiro, M. (2009). On some categories and functors in the theory of quaternionic Bergman spaces. *Advances in Applied Clifford Algebras, 19*, 325–338.
22. Hakuk, Y., & Reich, Y. (2023). Computational infrastructure for concepts discovery in science and technology. *Advanced Engineering Informatics, 56*, 101938.
23. Helemskii, A. Ya. (2005). *Lectures and exercises on functional analysis*. American Mathematical Society.
24. Hofmann, D., & Reis, C. D. (2013). Probabilistic metric spaces as enriched categories. *Fuzzy Sets and Systems, 210*, 1–21.
25. Jost, J. (2015). *Mathematical concepts*. Springer International Publishing.
26. Kovalyov, S. P. (2017). Methods of category theory in model-based systems engineering. *Informatics and Applications, 11*(3), 42–50.
27. Landry, E. (Ed.). (2017). *Categories for the working philosopher*. Oxford University Press.
28. Mabrok, M. A., & Ryan, M. J. (2017). Category theory as a formal mathematical foundation for model-based systems engineering. *Applied Mathematics & Information Sciences, 11*(1), 43–51.
29. Mac Lane, S. (1978). *Categories for the working mathematician*. Springer.
30. Marquis, J.-P. (2016). Stairway to heaven: The abstract method and levels of abstraction in mathematics. *The Mathematical Intelligencer, 38*(3), 41–51.
31. Mordecai, Y., & Engel, A. (2024). Systems engineering using category theory. In A. Engel (Ed.), *Systems science for engineers and scholars* (pp. 410–447). Wiley.
32. Pierce, B. C. (2002). *Types and programming languages*. MIT Press.
33. Raczynski, S. (2012). Semi-discrete events and models in categorical language. *International Journal of Simulation Modelling, 11*(2), 89–96.
34. Rine, D. C. (1973). A category theory for programming languages. *Mathematical Systems Theory, 7*, 304–317.
35. Sowa, J. (2000). *Knowledge representation: Logical, philosophical, and computational foundations*. Brooks/Cole.
36. Spivak, D. (2014). *Category theory for the sciences*. MIT Press.

37. Spivak, D., & Kent, R. (2012). Ologs: A categorical framework for knowledge representation. *PLoS One, 7*(1), e24274.
38. Spivak, D. I., & Fong, B. (2019). *An invitation to applied category theory. Seven sketches in compositionality*. Cambridge University Press.
39. Walters, R. F. C. (1991). *Categories and computer science*. Cambridge University Press.
40. Wisnevsky, R., Breiner, S., Jones, A., Spivak, D. I., & Subrahmanian, E. (2017). Using category theory to facilitate multiple manufacturing service database integration. *Journal of Computing and Information Science in Engineering, 17*(2), 021011 (11 pp.).

Chapter 3
Categorical Approach to Mathematical Modelling

Abstract In this chapter, we introduce a category theory-based conceptual modelling framework, that formally describes physics-based models appearing in engineering practice. At first, a categorical description of mathematical models is introduced, and their properties are discussed. After that, coupled models are addressed by lifting the results from categories of mathematical models to the categories of coupled models. However, because of the coupling, the structure of categories of coupled models is more complicated. In particular, it is necessary to distinguish if models from the same or from different categories are coupled. Finally, more abstract categorical constructions on categories of mathematical models are discussed. For supporting the reader, the chapter is provided with various illustrative examples for practical interpretations of the abstract constructions introduced in this chapter.

3.1 Preliminary Remarks

As discussed in Chap. 1, the complexity of modern engineering problems necessitates the use of advanced coupled models constructed as a combination of partial models describing each individual phenomenon needed to be addressed by the global model. The category theory-based modelling framework presented in this chapter aims to provide a consistent description of models and their couplings in the language of category theory. Abstract categorical descriptions require first the definition of *universal properties of models*—properties which are shared by models in general and are independent on any specific problem or engineering field. If a universal model property is defined, then all categorical constructions used in one specific modelling application can be directly transferred to another field. Thus, the category theory-based modelling framework begins with defining such a universal model property which is common across all models. Furthermore, the main goal is also to ensure that the constructions introduced in the framework retain meaningful physical and engineering interpretations, bridging the abstract theory with practical applications. Hence, we will provide illustrative examples alongside the abstract categorical constructions.

D. Legatiuk, *Mathematical Modelling by Help of Category Theory*,
Mathematical Engineering, https://doi.org/10.1007/978-3-031-84951-0_3

For fixing the setting, let us recall that, from the modelling perspective, models used in practice can be generally classified into two types:

- *physics-based models*—models which are based on mathematical formalisations of physical laws and assumptions;
- *data-driven models*—models which are based on representations of data, e.g. results of experiments or measurements obtained from a monitoring system.

In this chapter, we will refer to physics-based models simply as "mathematical models", since this type of models is typically implied when we speak about mathematical modelling. Furthermore, our primary focus will be on mathematical models, as they provide a richer basis for conceptual modelling compared to data-driven models. This distinction arises from the fact that individual components of mathematical models can often be interpreted from the modelling perspective, but individual components of data-driven models do not generally have any practical meaning: for example a term appearing in a regression formula has no particular practical meaning.

As it discussed in Chap. 1, the modern modelling process requires a structured approach that enables the clear identification of different critical factors influencing the quality of the final result. Therefore, the primary motivation for this chapter is to propose a modelling framework for the conceptual modelling phase based on category theory. This framework aims to provide a *structural view* on mathematical models and especially on the construction of coupled models that appear in engineering applications.

From the mathematical perspective, a "structural view" on a field requires selecting a formal language to describe relations between different objects of this field, i.e. defining a general structure of the field. In the context of modelling, this implies the need for a formal language that can effectively describe relations between different models. To elaborate on this point further, let us consider the following differential equations:

$$
\begin{aligned}
&\rho F \frac{\partial^2 u}{\partial t^2} + E I_y \frac{\partial^4 u}{\partial x^4} &&= 0,\\
&\rho F \frac{\partial^2 u}{\partial t^2} + E I_y \frac{\partial^4 u}{\partial x^4} - \rho I_y \frac{\partial^4 u}{\partial x^2 \partial t^2} &&= 0, \qquad (3.1)\\
&\rho F \frac{\partial^2 u}{\partial t^2} + E I_y \frac{\partial^4 u}{\partial x^4} - \rho I_y \left(1 + \frac{E}{\aleph \mu}\right) \frac{\partial^4 u}{\partial x^2 \partial t^2} + \frac{\rho^2 I_y}{\aleph \mu} \frac{\partial^4 u}{\partial t^4} &&= 0.
\end{aligned}
$$

It could be recognised that these differential equations corresponds to three common beam theories: the Bernoulli-Euler theory, the Rayleigh theory, and the Timoshenko theory, respectively. Thus, it can be concluded that these equations describe the transverse vibrations of a one-dimensional beam, and therefore, coefficients have precise physical meaning: ρ is the material density, F is the area of cross section, E is the Young modulus, I_y is the moment of inertia, μ is the shear modulus, κ is the Timoshenko shear coefficient, and the unknown function $u(x, t)$ represents the displacement of points of the beam.

Let us formulate now first *structural question*:

How the three models of transverse vibrations of one-dimensional beams given by (3.1) are related?

We can specify this question even more:

What makes the three models of transverse vibrations of one-dimensional beams given by (3.1) different?

Perhaps the most obvious simple answer to the last question is the form of equations in (3.1). Let us accept this answer for now and consider another example: the linear elasticity model, which describes the deformations of an elastic body in a static case, is formalised by the following equations

$$\begin{cases} \operatorname{div} \tilde{\boldsymbol{\sigma}} + \rho\, \mathbf{K} = 0, \\ \tilde{\boldsymbol{\varepsilon}} = \dfrac{1}{2}\left[\nabla \mathbf{u} + (\nabla \mathbf{u})^{\mathrm{T}}\right], \\ \tilde{\boldsymbol{\sigma}} = 2\mu \left(\dfrac{\nu}{1-2\nu} \vartheta\, \tilde{\mathbf{E}} + \tilde{\boldsymbol{\varepsilon}}\right), \end{cases} \qquad \vartheta = \operatorname{div} \mathbf{u} = \frac{\partial u_1}{\partial x_1} + \frac{\partial u_2}{\partial x_2} + \frac{\partial u_3}{\partial x_3}, \tag{3.2}$$

where $\tilde{\boldsymbol{\sigma}}$ is the symmetric stress tensor, $\tilde{\boldsymbol{\varepsilon}}$ is the symmetric strain tensor, $\mathbf{u}$ is the displacement vector, ρ is the material density, ν is the Poisson's ration, and $\mathbf{K}$ is the volume force. System of equations (3.2) is the classical tensor version of the elasticity equations, see for example [19]. However, the Lamé equation

$$\mu\, \Delta \mathbf{u} + (\lambda + \mu) \operatorname{grad} \operatorname{div} \mathbf{u} + \rho \mathbf{K} = 0, \tag{3.3}$$

is often used in practice as well. Furthermore, the model of linear elasticity can be also written as follows

$$DMDu = 0, \text{ with } D = \sum_{k=1}^{3} \mathbf{e}_k \partial_k, \text{ and } u = u_0 + \mathbf{u}, \tag{3.4}$$

where the multiplication operator M is defined by

$$M\mathbf{u} := \frac{m-2}{2(m-1)} u_0 + \mathbf{u}, \quad m := \nu^{-1}.$$

Equation (3.4) is a quaternionic form of the elasticity model with D denoting the Dirac operator, see [8] for details.

Based on this discussion, we can formulate now another structural question:

Obviously Eqs. (3.2)–(3.4) are different, but describing the same model of elasticity, and can be transformed one into another; can we formally describe this transformation, and how can we characterise models generally, i.e. without comparing final equations?

In this book, we primarily focus on the question of model comparability, and we will discuss only some special situations for the case when the same model is represented in different formulations. As we will see, the latter case is generally more involved, since different formulations of the same problem may arise not only by different ways of working with basic modelling/physical assumptions, but also from the fundamental differences in the *nature* of these derivations. If the case of a different treatment of the basic modelling assumptions will be discussed in this book, the scenario in which different modelling assumptions lead, under certain constraints, to the same models will be only mentioned briefly.

3.2 Mathematical Models as Categories

As discussed earlier, category theory will be employed in this chapter to develop a "structural view" on mathematical models. This categorical perspective will serve as the foundation for a conceptual modelling framework that addresses the coupling and formal comparison of models. The construction of category theory-based modelling framework can be approached from two different perspectives:

1. **Modelling perspective**: In this approach, all categorical constructions are initially motivated by the principles of mathematical modelling and are later formalised using the language of category theory.
2. **Abstract perspective**: In this approach, a few basic axioms are introduced in the beginning, and all subsequent constructions are derived solely by using the tools of category theory, without direct motivation from the modelling perspective.

Arguably, the abstract perspective is more "natural" from the perspective of constructing mathematical theories. However, the modelling perspective is more suitable for our context, because the final goal is to support practical mathematical modelling—a focus which might be lost while pursuing the beauty of abstract constructions. Therefore, the modelling perspective will dominate the construction of the category theory-based modelling framework presented in this chapter. This emphasis is reflected in the inclusion of numerous illustrative examples and clear modelling motivations for the specific categorical constructions chosen.

3.2.1 Categories of Mathematical Models

We begin our construction with the introduction of *concrete* categories $\mathbf{Model}_i$, $i = 1, 2, \ldots$, which are *associated* with mathematical models used to describe a certain physical phenomenon, such as, for example, models of elasticity theory or heat conduction. The term "associated" has been used, because, strictly speaking, the objects of categories $\mathbf{Model}_i$, $i = 1, 2, \ldots$ are not mathematical models themselves, but rather sets of basic physical assumptions upon which the corresponding mathematical models are created. However, to keep notations short and transparent, we will refer to these categories simply as *categories of mathematical models*.

The following definition introduces the basic structure of categories of mathematical models:

Definition 3.1 (Category of Mathematical Models) Let $\mathbf{Model}_1$ be a category of mathematical models describing a given physical phenomenon. Then for all objects of $\mathbf{Model}_1$ the following assumptions hold:

1. each object is a finite non-empty set—*set of assumptions* of a mathematical model, denoted by $\mathbf{Set}_{\mathrm{A}}$, where A is the corresponding mathematical model;
2. morphisms (arrows) are relations between these sets;
3. for each set of assumptions and its corresponding model exists a mapping

$$\mathbf{Set}_{\mathrm{A}} \overset{S}{\mapsto} \mathrm{A};$$

4. all objects are related to mathematical models acting in the same physical dimension.

This definition is motivated by a modelling perspective: models are not immediately given in terms of mathematical expressions and equations; instead, these expressions, which represent the models, are derived from basic physical assumptions. This perspective forms the basis for the four assumptions listed in the definition. To clarify the modelling perspective further, let us provide some comments on the assumptions used in this definition:

- **Assumption 1**. This assumption arises naturally from the modelling process discussed in Chap. 1. A mathematical model is created to describe a certain physical phenomenon or process, and this is possible only if the physical background of the phenomenon or process is clearly defined. In other words, the assumptions to be satisfied by the model need to be formulated in advance.

  Furthermore, to ensure a clear distinction between different mathematical models, the set of assumptions is understood in a broader sense: not only basic physical assumptions are listed, but also any subsequent modifications and simplifications of the model, such as for example, a linearisation of original equations.

The requirement for the set of assumptions to be finite comes from the fact that no model possess an infinite set of physical assumptions. Hence, considering more general sets is not necessary.

- **Assumption 2**. This assumption, in fact, introduces the structure of categories of mathematical models. The main point here is that instead of working with discrete categories, it is beneficial to study more elaborated structure. Because the objects in categories of mathematical models are sets, it is natural to use relations between sets as morphism in these categories. We will make these relations more specific and connect it with a clear practical interpretation in the next section.
- **Assumption 3**. This assumption formally describes the process of deriving the final form of a model, for example a differential or integral equation, from basic physical assumptions. In this context, the mapping S represents a *formalisation process* consisting in expressing the basic physical assumptions in terms of mathematical expressions that constitute the mathematical model. In fact, this formalisation process is the actual modelling process, or, better to say, the process of creating a model. Section 5.1 presents first ideas on using type theory to describe the formalisation process of physical assumptions. The main idea here is to detect modelling errors on the early stage of modelling process by utilising advantages of a properly designed type system.

  We also would like to remark, that originally, the mapping S has been called *invertible* in [9]. The invertibility in this case means, that the set of assumptions can be uniquely reconstructed from the final form of a model. Although, that such a reconstruction can be indeed possible for some simple models, this is not true in a general case. Even if we consider the following canonical parabolic equation

$$u_t = a^2 u_{xx},$$

  then without extra context it cannot be decided if this is a heat equation or a diffusion equation [22]. Thus, we prefer to use the formulation of Assumption 3 given in Definition 3.1.
- **Assumption 4**. This assumption is necessary to ensure that models from different dimensions are not placed in the same category. This distinction is essential because dimensional reduction often simplifies a model. These simplifications should be treated as modelling assumptions and listed in the sets of assumptions discussed above.

  Moreover, separation of models with respect to their dimensionality provides a clearer structure of categories and enhanced transferability. In the case of placing models from different dimensions in the same category, we would have then multiple models, which are almost identical, and, hence, making the structure of the category less clear. If, for instance, a category of mathematical models is created for the three-dimensional case, then its structure can be straightforwardly transferred to the two-dimensional case by fine-tuning the objects (sets of assumptions).

Additionally, Assumption 4 implies that models must describe a physical phenomenon in the same domain. For example, consider a heat conduction process in a domain $\Omega = \Omega_1 \cup \Omega_2$ made of two different material (subdomains Ω_1 and Ω_2). In this case, the models used in the subdomains must be treated separately, even though they technically belong to the same category. This case will be discussed in detail in Sect. 3.3.

Let us finish this section by providing some further remarks about the structure of categories of mathematical models:

Remark 3.1 It is necessary to comment that having finite sets as objects in the category is one possible approach to address mathematical models. In general, it is possible to think of categories of mathematical models where the objects are mathematical expressions (equations) that represent the models. However, in this case it will be more difficult to distinguish models, because the same set of assumptions can lead to different formalisations in terms of final equations, see examples in the beginning of this chapter, as well as the same equation can represent different physical phenomena, as discussed above. Nonetheless, perhaps there is a clever way of treating this case, which is still to be found.

Remark 3.2 Additionally, we would like to remark, that models with different parameters, such as material constants, can correspond to the same set of assumptions. For example, if we consider the set of assumptions that leads to the Lamé equation (a partial differential equation with constant coefficients), then it is clear that an infinite number of possible constant coefficients exists, but all these specific models originate from the same set of assumptions.

In general, models originating from the same set of assumptions, but having different material parameters, are just particular instances of this set of assumptions. In this way, it becomes evident that stochasticity only in material parameters does not change basic modelling assumptions, because the fact that a constant is chosen according to a certain probability law does not principally affect the assumption of having constant coefficients.

In contrast, modelling of physical process by help of stochastic partial differential equations is based on entirely different modelling assumptions, see for example [17], and therefore, should not be put together with "classical" mathematical models. Unfortunately, in modern engineering, the models with stochasticity only in material parameters are often also referred to as *stochastic models*. This terminology is not generally accurate due to the reasons discussed above.

In this work, we will always pursue a rigorous distinction between different types of models, and we will try to clarify possible sources of confusions regarding the engineering literature, where it is necessary.

Remark 3.3 We would like to remark that in works [8, 13] objects in categories $\mathbf{Model}_i$, $i = 1, 2, \ldots$ were typically directly referred to as mathematical models. This referencing of objects as mathematical models has been motivated by the modelling point of view and by the desire to simplify notations. However, from the theoretical point of view, this is of course not entirely correct, since objects in our categories are finite sets associated with mathematical models via a mapping S, see Definition 3.1 again, and not mathematical models directly. In this work, we would like to keep our constructions more rigorous, and therefore, we will try to always keep an explicit distinction between objects in categories $\mathbf{Model}_i$, $i = 1, 2, \ldots$ and the corresponding mathematical models, as we have already done in the above constructions.

3.2.2 Relations Between Mathematical Models

Definition 3.1 introduces the general structure of categories of mathematical models, where sets of assumptions serve as objects and relations between these sets act as morphism. The next step is to specify these relations within the modelling context. The main requirement for these relations is that they must define a *universal model property*—a property independent on any specific problem. This implies that boundary or initial conditions (but not coupling or transmission conditions) should not influence the model property. Therefore, to satisfy this restriction, the comparison of mathematical models using a relative model property, referred to as *model complexity*, is proposed:

Definition 3.2 (Complexity of Mathematical Models) Let A and B be mathematical models in a category $\mathbf{Model}_1$. We say that model A has higher complexity than model B if and only if $\mathbf{Set}_{\mathrm{A}} \subset \mathbf{Set}_{\mathrm{B}}$, but $\mathbf{Set}_{\mathrm{B}} \not\subset \mathbf{Set}_{\mathrm{A}}$. Consequently, two models are called equal, in the sense of complexity, iff $\mathbf{Set}_{\mathrm{B}} = \mathbf{Set}_{\mathrm{A}}$.

The model complexity in this definition is defined relatively, since we do not describe it explicitly. From the point of view of physics, model complexity reflects the fact that a model which has fewer assumptions provides a more accurate description of a physical phenomenon under consideration. Thus, the model complexity is a *relative quality measure* of how good a mathematical model represents a given physical phenomenon. The relativity in the measure comes from the fact, that any comparison needs at least two objects, and one model cannot be assessed with respect to its ability to represent the corresponding physical process, otherwise that would imply that the exact representation of the physical process is known *a priori*.

From a more general perspective, the model complexity introduced in Definition 3.2 can be seen as an abstract approach to formalise the concept of *model fidelity*—a measure of how well a model represents reality. We will discuss the connections between model fidelity and the abstract constructions introduced in the category theory-based modelling framework in Chap. 4.

It is necessary to remark that the notion of model complexity introduced above is neither related to the definition of complexity typically used in computer science (complexity of an algorithm), nor to the definition of complexity used for statistical model, where the number of parameters is typically served as a complexity measure. Thus, the notion of model complexity introduced in Definition 3.2 represents in a unique way the complexity of mathematical models in general, based on the difference in their sets of assumptions. Moreover, this notion of model complexity does not depend on specific boundary or initial conditions, since basic model assumptions are generally not influenced by specific boundary or initial conditions; however, if boundary conditions are essential for basic model assumptions, e.g. singular boundary conditions, then they will be automatically listed in the corresponding set of assumptions, since such boundary conditions are critical for describing the physical process. Thus, the model complexity introduced in Definition 3.2 is a universal model property.

It is also worth to discuss the point, that Definition 3.2 might sound a bit counter-intuitive, since it states that a model satisfying less modelling assumption is more complex, and not of higher *simplicity*, as it could be expected as well. In fact, both points of view on the complexity are possible, and differ only in the general understanding of modelling assumptions. Definition 3.2 is based on the idea that modelling assumptions act as *restrictions* for a model, and thus implying that a model with less modelling assumptions is more general, which is similar the lattice of theory context discussed in Sect. 2.3. Nonetheless, another perspective on the notion of model complexity still can be considered, which would reflect the opposite point of view: model assumptions are not restrictions, but rather *generalisations* of models. This discussion is also directly related to the following important remark:

Remark 3.4 Sets of assumptions introduced in Definition 3.1 are assumed to be written by the help of a natural language. While intuitively it is clear how to formulate these sets, as well as how to compare them in the sense of model complexity, from the formal perspective it is not so straightforward.

In fact, a formal comparison of sets of assumptions written in a natural language can be done only by the help of a detailed semantic analysis of these sentences, and only after that, sentences, and, consequently, sets of assumptions, can be rigorously compared. A possible way around this problem could be to impose stricter rules on formulating sets of assumptions. In that case, a kind of basic "alphabet" of allowed expressions and symbols could be introduced. Moreover, perhaps a combination of a natural language and mathematical expressions complemented by strict rules could be a suitable option. Although this topic goes beyond the scope of this book, we will provide remarks on processing the sets of assumptions in Chap. 5.

From the relational algebra perspective, the model complexity from Definition 3.2 is a binary relation in a category of mathematical models. Hence, the objects in categories of mathematical models can be ordered by using model complexity. However, the ordering of objects defined by model complexity is only partial, and not total, since examples of mathematical models which should belong to the same category but cannot be ordered according to Definition 3.2 can be easily found (see examples provided in Sect. 3.2.4). Naturally, in some cases, mathematical models can constitute a category with totally ordered objects. To have a clear distinction between categories with partial and total ordering of objects, we introduce the following definition:

Definition 3.3 Let $\mathbf{Model}_1$ be a category of mathematical models in which n objects $\mathbf{Set}_{A_j}$, $j = 1, \ldots, n$ can be ordered according to Definition 3.2 as follows

$$\mathbf{Set}_{A_i} \subset \mathbf{Set}_{A_j}, \quad \text{for } i < j \leq n.$$

Moreover, let X be the set of all modelling assumptions used in this category. Then category $\mathbf{Model}_1$ contains totally ordered objects, and therefore is associated with totally ordered models, if and only if

$$X = \mathbf{Set}_{\mathrm{A}_1} \cup \mathbf{Set}_{\mathrm{A}_2} \cup \ldots \cup \mathbf{Set}_{\mathrm{A}_n}, \quad \text{and } \mathbf{Set}_{\mathrm{A}_n} = X,$$

otherwise, the category $\mathbf{Model}_1$ contains partially ordered objects corresponding to partially ordered models.

As a direct consequence of this definition, we have the following corollary:

Corollary 3.1 *In a totally ordered category* $\mathbf{Model}_1$ *with n objects always exist two unique objects:*

- *object* $\mathbf{Set}_{\mathrm{A}_1}$ *satisfying* $\mathbf{Set}_{\mathrm{A}_1} \subset \mathbf{Set}_{\mathrm{A}_i}$ $\forall i = 2, \ldots, n$, *which is called* ***the most complex object****, and the associated model* A_1 *is called the most complex model;*
- *object* $\mathbf{Set}_{\mathrm{A}_n}$ *satisfying* $\mathbf{Set}_{\mathrm{A}_n} = \mathbf{Set}_{\mathrm{A}_1} \cup \mathbf{Set}_{\mathrm{A}_2} \cup \ldots \cup \mathbf{Set}_{\mathrm{A}_n}$, *which is called* ***the simplest object****, and the associated model* A_n *is called the simplest model.*

Let us make the following remark:

Remark 3.5 It is worth to mention, that in the framework of the modelling formalism introduced in this chapter, the most complex object and the simplest object are, in fact, the *initial object* and the *terminal object* in categories of mathematical models, respectively. Note that, although categories of mathematical models have finite sets as objects, the initial and terminal objects are different to the ones in the classical category **Sets**, where these are given by the empty set and one-element set, correspondingly. The difference comes precisely from the modelling background of our categories, since while formally it is still possible to consider the empty and one-element sets as sets of assumptions of some (fictitious) models, it does not make sense from the modelling perspective.

The proof of Corollary 3.1 is straightforward, and we only would like to mention, that uniqueness of objects $\mathbf{Set}_{\mathrm{A}_1}$ and $\mathbf{Set}_{\mathrm{A}_n}$ follows immediately from Definition 3.2 and from the fact that a totally ordered category is considered. The situation is more tricky in the case of partially ordered categories, as shown in the following proposition:

Proposition 3.1 *For a partially ordered category* $\mathbf{Model}_1$ *with n objects one of the following statements holds:*

1. *the most complex object* $\mathbf{Set}_{A_1}$ *and the simplest object* $\mathbf{Set}_{A_n}$ *do not exist;*
2. *the most complex object* $\mathbf{Set}_{A_1}$ *exists, while the simplest object* $\mathbf{Set}_{A_n}$ *does not exist;*
3. *the most complex object* $\mathbf{Set}_{A_1}$ *does not exist, while the simplest object* $\mathbf{Set}_{A_n}$ *exists;*
4. *the most complex object* $\mathbf{Set}_{A_1}$ *and the simplest object* $\mathbf{Set}_{A_n}$ *exist simultaneously.*

Proof We prove this proposition by straightforwardly constructing corresponding structures of partially ordered categories. Modelling background of this construction will become clear from practical examples presented later on in this chapter.

We start the proof by proving cases 2 and 3 at first, since the proof of case 1 will be based on cases 2 and 3. We consider a category with one object $\mathbf{Set}_{A_1}$, and the rest objects we construct explicitly from $\mathbf{Set}_{A_1}$. Without loss of generality, we assume that $\mathbf{Set}_{A_1}$ contains at least one element, which will be denoted by $A_1^{(1)}$. The objects $\mathbf{Set}_{A_2}$ and $\mathbf{Set}_{A_3}$ are then constructed from $\mathbf{Set}_{A_1}$ by adding different elements $A_1^{(2)}$ and $A_1^{(3)}$ to $\mathbf{Set}_{A_1}$, correspondingly. Consequently, we obtain new sets of assumptions by adding two different assumptions. This construction is shown by the diagram

$$\begin{array}{ccc} & \mathbf{Set}_{A_1} = \left\{A_1^{(1)}\right\} & \\ \swarrow & & \searrow \\ \left\{A_1^{(1)}, A_1^{(2)}\right\} = \mathbf{Set}_{A_2} & & \mathbf{Set}_{A_3} = \left\{A_1^{(1)}, A_1^{(3)}\right\} \end{array}$$

implying that $\mathbf{Set}_{A_1} \subset \mathbf{Set}_{A_2}$ and $\mathbf{Set}_{A_1} \subset \mathbf{Set}_{A_3}$, but $\mathbf{Set}_{A_2}$ and $\mathbf{Set}_{A_2}$ are not related. Thus, $\mathbf{Set}_{A_1}$ is the most complex object in this category, but the simplest object does not exists. Hence, the case 2 is proved.

The proof of case 3 is analogues to case 2, where only instead of adding extra assumptions, we remove different assumptions from the initial set. Thus, for simplicity, we assume that $\mathbf{Set}_{A_1}$ has at least two different assumptions. The rest of proof follows immediately.

To prove case 1, we consider now two distinct objects $\mathbf{Set}_{A_1}$ and $\mathbf{Set}_{A_2}$ given by

$$\mathbf{Set}_{A_1} = \left\{A_1^{(1)}, A_1^{(2)}\right\} \text{ and } \mathbf{Set}_{A_2} = \left\{A_1^{(1)}, A_1^{(2)}, A_2^{(1)}\right\},$$

respectively. Similar to cases 2 and 3, we construct now two other objects in two different ways, as follows:

$$\mathbf{Set}_{A_3} = \left\{A_1^{(1)}, A_1^{(2)}\right\} \setminus \left\{A_1^{(2)}\right\} = \left\{A_1^{(1)}\right\},$$

$$\mathbf{Set}_{A_4} = \left\{A_1^{(1)}, A_1^{(2)}\right\} \setminus \left\{A_1^{(1)}\right\} = \left\{A_1^{(2)}\right\},$$

and

$$\mathbf{Set}_{A_3} = \left\{A_1^{(1)}, A_1^{(2)}, A_2^{(1)}\right\} \setminus \left\{A_1^{(2)}, A_2^{(1)}\right\} = \left\{A_1^{(1)}\right\},$$

$$\mathbf{Set}_{A_4} = \left\{A_1^{(1)}, A_1^{(2)}, A_2^{(1)}\right\} \setminus \left\{A_1^{(1)}, A_2^{(1)}\right\} = \left\{A_1^{(2)}\right\}.$$

This construction is illustrated by the following diagram:

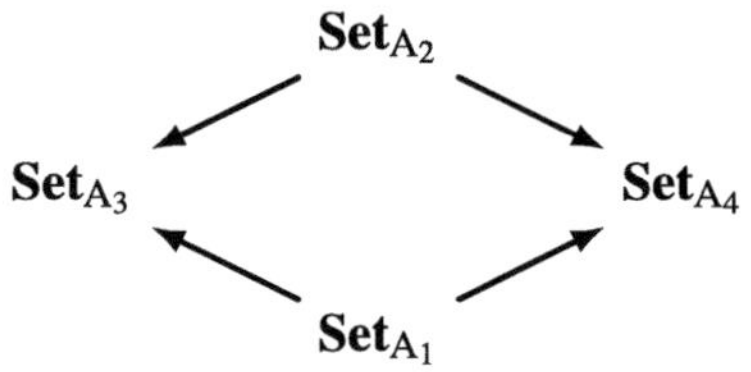

Thus, the constructed category is partially ordered, and since objects $\mathbf{Set}_{A_1}$ and $\mathbf{Set}_{A_2}$ are not related, this category contains neither the most complex nor the simplest objects, since no object satisfies assumptions of Corollary 3.1.

For proving case 4, let us consider the object $\mathbf{Set}_{A_1} = \left\{A_1^{(1)}, A_1^{(2)}, A_1^{(3)}, A_1^{(4)}\right\}$, and let us construct several other objects according to the following commutative diagram

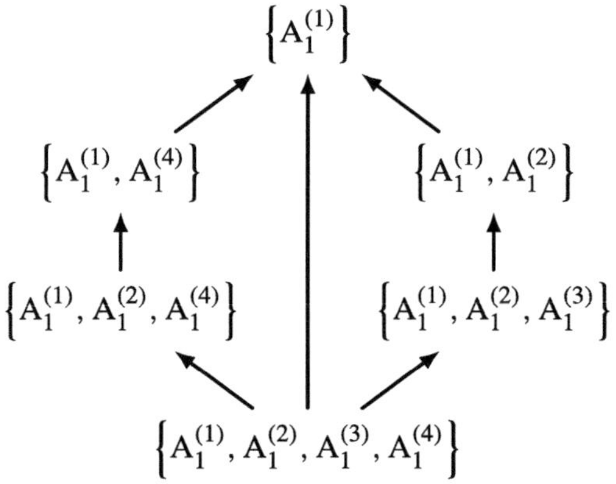

Although the diagram is commutative, but the objects on the left side are not related to the objects of the right side in the sense of Definition 3.2. Thus, we have a partially ordered category, where both the most complex object $\left\{A_1^{(1)}\right\}$ and the simplest object $\left\{A_1^{(1)}, A_1^{(2)}, A_1^{(3)}, A_1^{(4)}\right\}$ exist simultaneously. Hence, the proposition is proved. □

Finally, we summarise our discussion in the form of the following theorem:

Theorem 3.1 *Consider a category* $\mathbf{Model}_1$ *with* n *objects. If the most complex object* $\mathbf{Set}_{\mathrm{A}_1}$ *and the simplest object* $\mathbf{Set}_{\mathrm{A}_n}$ *exist simultaneously in the category* $\mathbf{Model}_1$*, then* $\mathbf{Model}_1$ *is either a totally ordered category, or contains at least two totally ordered subcategories.*

Proof The proof of this theorem follows immediately from Corollary 3.1, Proposition 3.1, and Definition 3.3. Looking at the proof of the case 4 in Proposition 3.1, we see immediately that two totally ordered subcategories exist. The case of only one totally ordered subcategory is excluded by the assumption of simultaneous existence of the most complex and the simplest objects. Further, if the most complex and the simplest objects exist simultaneously and all objects in the category $\mathbf{Model}_1$ are related by help of complexity, then it follows immediately that $\mathbf{Model}_1$ is a totally ordered category. □

To finalise our discussion on partially and totally ordered categories of mathematical models, we need to establish a connection between the both. Considering Definition 2.10, it becomes evident, that a totally ordered category can be obtained from a partially ordered category by "leaving out" objects, which cannot be compared by the complexity definition. In this case, the totally ordered category is a subcategory of the partially ordered category, where some objects have been removed. Such a subcategory contains at least two objects, since even in the case of a partially ordered category at least two objects are related in the sense of Definition 3.2, otherwise these objects would not constitute a category of mathematical models.

Moreover, a partially ordered category can contain several totally ordered subcategories. For instance, consider the following trivial example of a partially ordered category with three objects:

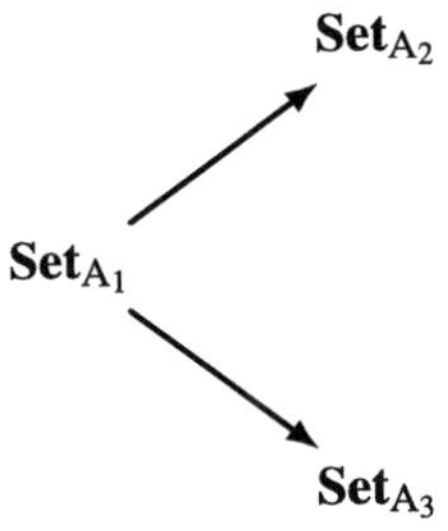

This category contains two totally ordered categories with objects $\{\mathbf{Set}_{\mathrm{A}_1}, \mathbf{Set}_{\mathrm{A}_2}\}$ and $\{\mathbf{Set}_{\mathrm{A}_1}, \mathbf{Set}_{\mathrm{A}_3}\}$, respectively.

Finally, the above discussion on the connection between partially and totally ordered categories can be summarised by the following theorem:

Theorem 3.2 *Every partially ordered category of mathematical models contains at least one totally ordered category of mathematical models as a subcategory.*

3.2.3 Convertible Mathematical Models

Let us now discuss the mappings between the sets of assumptions and the corresponding models in more detail. As it has been mentioned in the discussion of Definition 3.1, these mappings are generally not invertible because they represent a formalisation process of basic modelling assumptions in terms of mathematical expressions. Furthermore, these mappings are also not unique, as the same set of assumptions can be formalised differently. However, if objects in a category have been ordered (either partially or totally) based on their complexity, then these mappings will preserve this structure. Thus, these mappings are structure-preserving, meaning they are functors.

Because the mappings between sets of assumptions and the corresponding mathematical models are functorial, then, in fact, the mathematical models themselves also form a category. However, because the final form of a model depends on the formalisation process, it is more difficult to work directly with categories of models, rather than to describe categories of sets of assumptions, as we have done earlier. Nonetheless, we will now highlight some results related to the models directly. First, we summarise the preceding discussion in the following definition:

Definition 3.4 Let $\mathbf{Set}_{\mathrm{A}_1}$ be an object in the category $\mathbf{Model}_1$, and let B_1 and B_2 be two possible model formulations associated with the object $\mathbf{Set}_{\mathrm{A}_1}$ via two functors F and G. Then the model formulations B_1 and B_2 are connected via a natural transformation of functors ϑ, and the model formulations B_1 and B_2 are called *convertible*. This construction corresponds to the commutative diagram

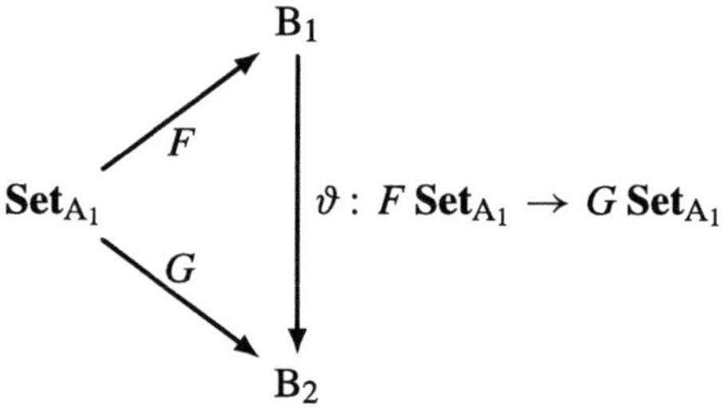

Moreover, models which are instantiated by convertible model formulations will be called *convertible models*.

Obviously, because different model formulations are related to the same set of assumptions, the model complexity of these formulations remains the same. Consequently, the following immediate corollary can be straightforwardly formulated based on Definition 3.4:

Corollary 3.2 *Convertible models have the same complexity.*

The notion of convertible mathematical models clarifies the discussion around different formulations of elasticity models (3.2)–(3.4) presented in the beginning of this chapter. In this way, all three model formulations, namely classical tensor form (3.2), the Lamé equation (3.3), and quaternionic form (3.4), do not constitute different models, but represent only different formalisations of the same set of assumptions.

Let us make two important remarks:

Remark 3.6 The discussion regarding convertible mathematical models further emphasises why sets of assumptions are considered as objects in categories of mathematical models, and not the model formulations directly. To illustrate this point, let us assume for a moment the opposite—that model formulations are treated as objects—and consider the following diagram with three objects for simplicity:

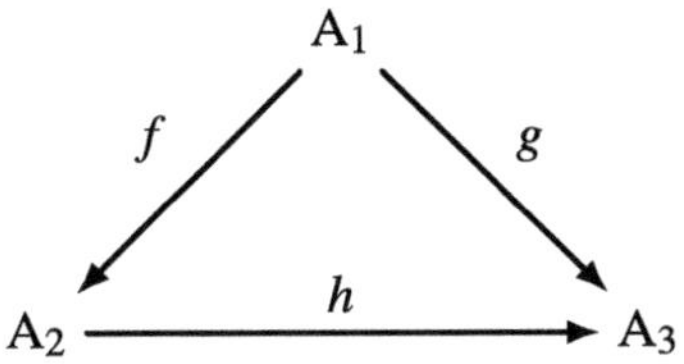

Moreover, assume additionally that the model formulations A_1 and A_2 are convertible in the sense of Definition 3.4, while the model formulation A_3 is not associated with the same set of assumptions. Thus, we would end up with two kinds of morphisms in the category: morphism f plays the same role as the natural transformation ϑ in Definition 3.4, while morphisms g and h represent complexity-relation on the level of model formulations. Obviously, it is necessary to be able to distinguish between the two kinds of morphisms, which would imply much more complicated constructions for the structure of the category, as well as for the relations between its objects.

Remark 3.7 Let us consider again the example with the elasticity models, and for the sake of readability, let us recall the three different model formulations from the beginning of the chapter:

$$
\mathrm{B}_1 := \begin{cases} \operatorname{div} \tilde{\boldsymbol{\sigma}} + \rho\, \mathbf{K} = 0, \\ \tilde{\boldsymbol{\varepsilon}} = \dfrac{1}{2}\left[\nabla \mathbf{u} + (\nabla \mathbf{u})^{\mathrm{T}}\right], \\ \tilde{\boldsymbol{\sigma}} = 2\mu \left(\dfrac{\nu}{1-2\nu} \vartheta\, \tilde{\mathbf{E}} + \tilde{\boldsymbol{\varepsilon}}\right), \end{cases} \qquad \vartheta = \operatorname{div} \mathbf{u} = \frac{\partial u_1}{\partial x_1} + \frac{\partial u_2}{\partial x_2} + \frac{\partial u_3}{\partial x_3},
$$

$$
\mathrm{B}_2 := \mu\, \Delta \mathbf{u} + (\lambda + \mu) \operatorname{grad} \operatorname{div} \mathbf{u} + \rho \mathbf{K} = 0,
$$

$$
\mathrm{B}_3 := DMDu = 0, \text{ with } D = \sum_{k=1}^{3} \mathbf{e}_k \partial_k, \text{ and } u = u_0 + \mathbf{u}.
$$

One could argue, that perhaps these model formulations should be correctly described by the diagram

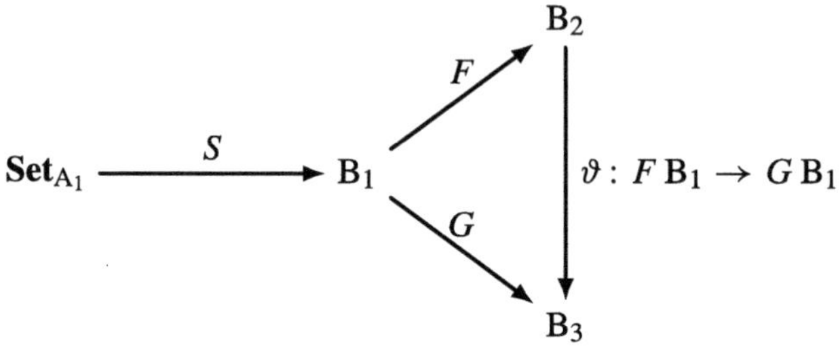

Here, functor S is a formalisation process of basic set of assumptions of linear elasticity $\mathbf{Set}_{\mathrm{A}_1}$ in the tensor form of model formulation B_1, after that, the tensor form can be further reformulated into the Lamé equation B_2, or into the quaternionic form B_3 via functorial mappings F and G. In some sense, the above diagram reflects the traditional way of developing different model formulations: at first, one original form is introduced, and after that, several more specific forms better suitable for specific problems or methods are introduced. Moreover, looking in particular at the quaternionic formulation B_3, it becomes clear that this form is not obtained directly through the formalisation process of $\mathbf{Set}_{\mathrm{A}_1}$ (at least no quaternionic-based modelling of linear elasticity has been reported till now), but through reformulation of either Lamé equation or the tensor form, see for example [2, 7] or more specific [3]. This is true not only for spatial hypercomplex formulations, but also for nowadays classical two-dimensional complex formulations, as provided by famous Kolosov-Muskhelishvili formulae [20]:

$$\begin{aligned} 2\mu(u_1 + i\,u_2) &= \kappa\,\Phi(z) - z\,\overline{\Phi'(z)} - \overline{\Psi(z)}, \\ \sigma_{11} + \sigma_{22} &= 2\left[\Phi'(z) + \overline{\Phi'(z)}\right] = 4\,\mathbf{Re}\left[\Phi'(z)\right], \\ \sigma_{22} - \sigma_{11} + 2i\,\sigma_{12} &= 2[\bar{z}\,\Phi''(z) + \Psi'(z)], \end{aligned}$$

which describe the components of stress tensor and displacement vector in terms of two holomorphic function $\Phi(z)$ and $\Psi(z)$ of the complex variable z.

Although it seems that the distinction made by the previous diagram is arguably the best way to deal with convertible mathematical models, we nonetheless will continue to work in the sense of Definition 3.4 for all model formulations because of the following reasons:

- Obtaining the Lamé equation or the quaternionic form from the tensor formulation does not change the fact, that both of these model formulations are still associated with the set of assumptions of linear elasticity.
- Even if there are currently no reports on a direct quaternionic formalisation of the basic set of assumptions of linear elasticity, it does not mean that this formalisation cannot be done principally. If this formalisation is achieved, then it

would be necessary to restructure the diagrams and the corresponding categorical structures.

- Finally, for many models used in engineering applications, it is often almost impossible to find the thin line between the very first model formulation and subsequent formulations. In most case this distinction will be purely formal, as history of science has shown that many concepts were often introduced independently and in parallel by different researchers.

3.2.4 Some Illustrative Examples

Before discussing the categorical perspective on coupled mathematical models, let us provide some illustrative examples of using category theory-based modelling framework to describe relations between different models used in engineering. For these examples, we will try to establish clear links with the notions and definitions introduced in the previous sections. We start with the example of structuring beam theories, and then proceed with a more complicated example from the field of elasticity theory.

3.2.4.1 Categorical Modelling of Beam Theories

As it has been mentioned already, typically, transverse vibrations of one-dimensional beams are modelled by using one of three common beam theories: the Bernoulli-Euler theory, the Rayleigh theory, and the Timoshenko theory. For providing a clear overview of the categorical constructions introduced in the previous section, we will not only describe the structure of category of beam models, but also provide some comments on the modelling process. Hence, the process of formalisation of the sets of assumptions in terms of final equations describing transverse vibrations of one-dimensional beams will be highlighted.

Let us now consider a category of mathematical models, denoted by **Beam**, containing as objects sets of assumptions $\mathbf{Set}_{\mathrm{B-E}}$, $\mathbf{Set}_{\mathrm{R}}$, $\mathbf{Set}_{\mathrm{T}}$ corresponding to the Bernoulli-Euler, Rayleigh, and Timoshenko beam theories, respectively. We start our discussion on the construction of category **Beam** by explicitly listing the sets of assumptions, which are given in Table 3.1.

Remark 3.8 The assumptions, as listed in Table 3.1, are formulated by the help of natural language, however in some cases it is more convenient to formulate sets of assumptions directly in terms of mathematical expressions, or as a mixture of both. Although from the set-theoretic point of view such freedom in writing sets of assumptions is not completely justified, it is acceptable in our setting because each set of assumption written in natural language can be rigorously formalised in terms of mathematical expressions. Thus, writing mathematical expressions in sets of assumptions can be considered as a kind of *syntactic sugar*, similar to programming

Table 3.1 Sets of assumptions of beam theories

	Assumptions	$\mathbf{Set}_{\mathrm{B-E}}$	$\mathbf{Set}_{\mathrm{R}}$	$\mathbf{Set}_{\mathrm{T}}$
1.	Cross sections of a beam that are planes remain planes after the deformation process	+	+	+
2.	Normal stresses on planes parallel to the axis of a beam are infinitesimal	+	+	+
3.	A beam has a constant cross section	+	+	+
4.	A beam is made of a homogeneous isotropic material	+	+	+
5.	Cross sections of a beam perpendicular to its axis remain perpendicular to the deformed axis	+	+	
6.	Rotation inertia of cross sections of a beam is omitted	+		

languages terminology. Of course, this analogy not perfect, but reflects a general point of view on writing sets of assumptions.

For a brief description of the modelling process, illustrating how different sets of assumptions listed in Table 3.1 are formalised in terms of partial differential equations, we consider the bending of a beam with a rectangular cross section. Among different possibilities to model beam equations known from mechanics, see for example [23], we will follow ideas from [5] (omitting some details). On this way, we will illustrate the modelling process for beam equations using Hamilton's principle, which require expressions for both potential and kinetic energies.

To find the potential energy, the work performed during the deformation process of the beam longitudinal symmetry axis needs to be expressed and then integrated with respect to the area of cross section. Set of assumption $\mathbf{Set}_{\mathrm{B-E}}$ of the Bernoulli-Euler theory leads to the following expression for the work of elastic forces

$$A = \frac{E}{2} z^2 \frac{\partial^2 u}{\partial x^2} b \, dz \, dx,$$

where E is the Young's modulus of the material, b is the width of cross section, and x, z are coordinates. Integrating this expression with respect to z, the density of potential energy is obtained in the form

$$dW_p = \frac{1}{2} E \frac{\partial^2 u}{\partial x^2} b \, dx \int_{-\frac{a}{2}}^{\frac{a}{2}} z^2 dz = \frac{1}{2} E I_y \frac{\partial^2 u}{\partial x^2} \, dx,$$

where $I_y = \dfrac{ba^3}{12}$ is the moment of inertia with a denoting the height of cross section. The density of kinetic energy is given by

$$dW_k = \frac{1}{2} \rho F \left(\frac{\partial u}{\partial t} \right)^2 dx,$$

where ρ is the density of material, and F is the area of cross section. Finally, by help of the Lagrangian $L = W_k - W_p$ and by using the Hamilton's variational principle, the beam equation in the framework of Bernoulli-Euler theory is obtained as follows:

$$\rho F \frac{\partial^2 u}{\partial t^2} + E I_y \frac{\partial^4 u}{\partial x^4} = 0.$$

If the assumptions of the Bernoulli-Euler theory are relaxed to account for the rotational inertia of the cross section, then the density of kinetic energy takes the following form

$$d W_k = \frac{1}{2} \left[\rho F \left(\frac{\partial u}{\partial t} \right)^2 + \rho I_y \left(\frac{\partial^2 u}{\partial x \partial t} \right)^2 \right] dx.$$

Again, by using the Lagrangian and the Hamilton's variational principle, the beam equation in the framework of Rayleigh theory is obtained

$$\rho F \frac{\partial^2 u}{\partial t^2} + E I_y \frac{\partial^4 u}{\partial x^4} - \rho I_y \frac{\partial^4 u}{\partial x^2 \partial t^2} = 0.$$

Finally, if the effect of bending of the cross sections is taken into account, then expressions for density of both potential and kinetic energies need to be modified. The new expressions are given by

$$d W_k = \frac{1}{2} \left[\rho F \left(\frac{\partial u}{\partial t} \right)^2 + \rho I_y \left(\frac{\partial \varphi}{\partial t} \right)^2 \right] dx,$$

$$d W_p = \frac{1}{2} \left[E I_y \left(\frac{\partial \varphi}{\partial x} \right)^2 + \aleph \mu F \left(\frac{\partial u}{\partial x} - \varphi \right)^2 \right] dx,$$

where φ is the angle of rotation of the normal to the mid-surface of the beam, $\aleph$ is the Timoshenko shear coefficient, which depends on the geometry of the beam, and μ is the shear modulus. Because these expressions have now two unknowns, namely the angle φ and the displacement u, in general, the Timoshenko beam theory corresponds to a system of two partial differential equations:

$$\begin{cases} \rho F \dfrac{\partial^2 u}{\partial t^2} - \aleph \mu F \dfrac{\partial^2 u}{\partial x^2} + \aleph \mu F \dfrac{\partial \varphi}{\partial x} = 0, \\[2ex] \rho I_y \dfrac{\partial^2 \varphi}{\partial t^2} - E I_y \dfrac{\partial^2 \varphi}{\partial x^2} + \aleph \mu F \left(\varphi - \dfrac{\partial u}{\partial x} \right) = 0. \end{cases}$$

After some calculations, this system can be reformulated in terms of only one partial differential equation for u as follows:

$$\rho F \frac{\partial^2 u}{\partial t^2} + E I_y \frac{\partial^4 u}{\partial x^4} - \rho I_y \left(1 + \frac{E}{\aleph \mu}\right) \frac{\partial^4 u}{\partial x^2 \partial t^2} + \frac{\rho^2 I_y}{\aleph \mu} \frac{\partial^4 u}{\partial t^4} = 0.$$

From a categorical perspective, the beam models and their corresponding sets of assumptions can be summarised as follows:

$$\mathbf{Set}_{\mathrm{B-E}} \overset{S}{\mapsto} \rho F \frac{\partial^2 u}{\partial t^2} + E I_y \frac{\partial^4 u}{\partial x^4} = 0 \qquad =: \mathbf{A},$$

$$\mathbf{Set}_{\mathrm{R}} \overset{S}{\mapsto} \rho F \frac{\partial^2 u}{\partial t^2} + E I_y \frac{\partial^4 u}{\partial x^4} - \rho I_y \frac{\partial^4 u}{\partial x^2 \partial t^2} = 0 \qquad =: \mathbf{B},$$

$$\mathbf{Set}_{\mathrm{T}} \overset{S}{\mapsto} \rho F \frac{\partial^2 u}{\partial t^2} + E I_y \frac{\partial^4 u}{\partial x^4} - \rho I_y \left(1 + \frac{E}{\aleph \mu}\right) \frac{\partial^4 u}{\partial x^2 \partial t^2} + \frac{\rho^2 I_y}{\aleph \mu} \frac{\partial^4 u}{\partial t^4} = 0 =: \mathbf{C}_1,$$

$$\mathbf{Set}_{\mathrm{T}} \overset{S}{\mapsto} \begin{cases} \rho F \dfrac{\partial^2 u}{\partial t^2} - \aleph \mu F \dfrac{\partial^2 u}{\partial x^2} + \aleph \mu F \dfrac{\partial \varphi}{\partial x} = 0, \\ \rho I_y \dfrac{\partial^2 \varphi}{\partial t^2} - E I_y \dfrac{\partial^2 \varphi}{\partial x^2} + \aleph \mu F \left(\varphi - \dfrac{\partial u}{\partial x}\right) = 0. \end{cases} \qquad =: \mathbf{C}_2,$$

where S are formalisation mappings, as discussed above. It is worth making the remark:

Remark 3.9 Note that, in general, mappings S can be different for each set of assumptions, or, can be the same as in the current example where they indicate the use of Hamilton's principle. If the fact that different formalisation processes have been used to obtain models from the sets of assumptions in one category is essential for the application, then it is necessary to indicate this fact by using sub-scripts such as $S_1, S_2, \ldots,$ otherwise the general notation for the formalisation mappings might be kept.

By using Definition 3.2, the category **Beam** can be straightforwardly equipped with the commutative diagram

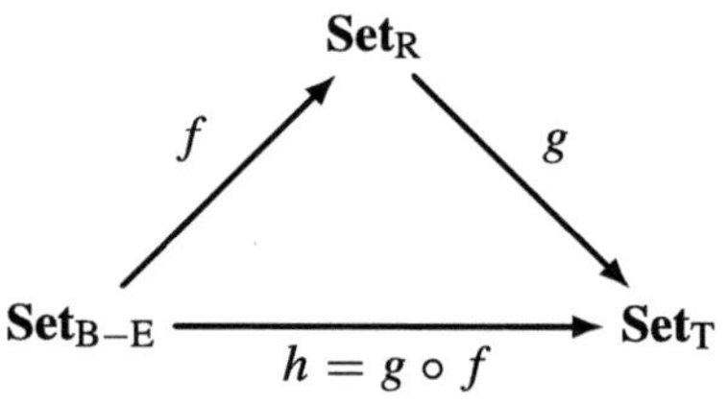

The morphisms f, g, and h indicate the simple fact, that one beam theory can be obtained from another by weakening basic assumptions. Furthermore, the above diagram clearly indicates that the object $\mathbf{Set}_{\mathrm{T}}$ (the Timoshenko theory) is the most complex, the object $\mathbf{Set}_{\mathrm{R}}$ (the Rayleigh theory) has higher complexity than the object $\mathbf{Set}_{\mathrm{B-E}}$ (the Bernoulli-Euler theory), which is the simplest object. The same ordering holds for the corresponding model formulations in terms of partial differential equations.

Next, let us list the following facts we know about the category **Beam**:

- **Beam** is a totally ordered category;
- The object $\mathbf{Set}_{\mathrm{B-E}}$ is the initial object of this category;
- The object $\mathbf{Set}_{\mathrm{T}}$ is the terminal object of this category;
- Models $\mathbf{C}_1$ and $\mathbf{C}_2$ are convertible, since they represent different formulations of the set of assumptions of the Timoshenko theory.

Note that, the first three facts, as well as the commutative diagram presented above, do not require, in fact, models themself, because these facts are solely obtained simply from the sets of assumptions, i.e. by analysing the objects in category **Beam**. Thus, the categorical point of view introduced in the previous section reflects the following idea:

The principle difference between models lies not in their final form, but in the basic modelling assumptions these models are constructed from.

Although this idea is very basic and perhaps obvious to most experienced modellers, it is surprisingly often overlooked in modern engineering.

Finally, at the level of models corresponding to the objects in **Beam**, we obtain the following diagram

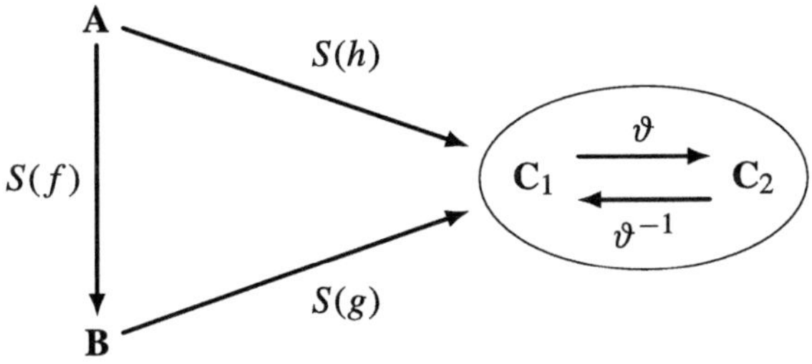

where ϑ denotes a natural transformation appearing in the definition of convertible models, recall Definition 3.4.

3.2.4.2 Categorical Modelling of Elasticity Theories

In this example, we will describe several models of elasticity theory, namely, classical linear elasticity [19], micropolar elasticity [4], linear elastic fracture mechanics [16], and nonlinear elasticity [18]. Consequently, we will introduce a category of mathematical models, denoted by **Elast**, containing as objects sets of assumptions $\mathbf{Set}_{\mathrm{LE}}$, $\mathbf{Set}_{\mathrm{ME}}$, $\mathbf{Set}_{\mathrm{LEF}}$, $\mathbf{Set}_{\mathrm{NE}}$ corresponding to the classical linear elasticity, micropolar elasticity, linear elastic fracture mechanics, and nonlinear elasticity, respectively. Unlike the beam models discussed previously, we will not describe in all details the derivation of various mechanical models. Instead, we briefly explain their basic modelling assumptions and model formulations. Furthermore, because various nonlinear elasticity models exist, for keeping the presentation brief, we will focus on the finite strains model. This excludes consideration of nonhomogeneous and plastic materials or, in other terms, physically nonlinear models.

Similar to the example of categorical modelling of beam theories, we start our discussion by explicitly listing the sets of assumptions (in a natural language) in Table 3.2.

Before presenting the final forms of mathematical expressions formalising the sets of assumptions listed in Table 3.2, let us provide at first a diagram to clarify the structure of the category **Elast**:

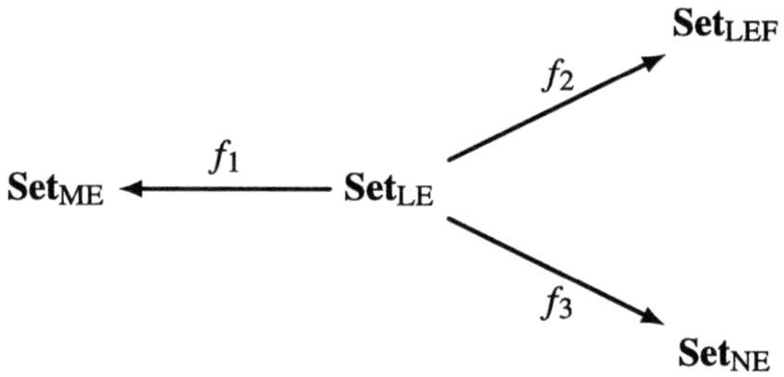

In contrast to the category **Beam**, the category **Elast** is a partially ordered category. Again, the morphisms f_1, f_2, and f_2 indicate that one elasticity theory can

Table 3.2 Sets of assumptions of elasticity theories

	Assumptions	$\mathbf{Set}_{\mathrm{LE}}$	$\mathbf{Set}_{\mathrm{ME}}$	$\mathbf{Set}_{\mathrm{LEF}}$	$\mathbf{Set}_{\mathrm{NE}}$
1.	A body is made of a homogeneous linear elastic material	+	+	+	+
2.	Strains appearing in a body are infinitesimal	+	+	+	
3.	Relationships between the components of stress and strain tensors are linear	+	+	+	
4.	Stress state does not produce yielding of a material	+	+	+	+
5.	A body does not contain cracks	+	+		+
6.	Microrotations of a continuum are not considered	+		+	+

be obtained from another by weakening basic modelling assumptions. Moreover, the above diagram clearly indicates that the object $\mathbf{Set}_{\mathrm{LE}}$ (classical linear elasticity theory) is the simplest object, while the most complex object does not exist in this category.

Let us list the following facts we know about the category **Elast**:

- **Elast** is a partially ordered category;
- The object $\mathbf{Set}_{\mathrm{LE}}$ is the initial object of this category;
- A terminal object does not exist in this category;
- **Elast** contains three totally ordered categories as sub-categories, each containing only two objects, precisely

$$\mathbf{Set}_{\mathrm{ME}} \xleftarrow{f_1} \mathbf{Set}_{\mathrm{LE}} \qquad \begin{array}{c} \mathbf{Set}_{\mathrm{LE}} \xrightarrow{f_2} \mathbf{Set}_{\mathrm{LEF}} \\ \mathbf{Set}_{\mathrm{LE}} \xrightarrow{f_3} \mathbf{Set}_{\mathrm{NE}} \end{array}$$

- Further, **Elast** contains three partially ordered categories as sub-categories with non-commutative structure, precisely

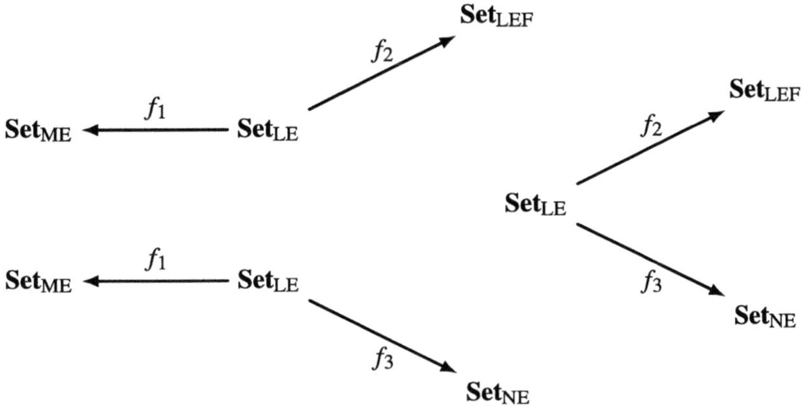

Next, we will present the model formulations arising from the sets of assumptions listed in Table 3.2. For simplicity, we focus only on the static case, although the dynamic case is also fully covered by our discussion.

The model of linear elasticity, derived from the object $\mathbf{Set}_{\mathrm{LE}}$, is typically formalised by the set of three equations, see for example [19, 23] for details:

1. Constitutive equations

$$\sigma_{ij} = \lambda \delta_{ij} \varepsilon_{kk} + 2\mu \varepsilon_{ij}, \quad i, j, k = 1, 2, 3,$$

where σ_{ij} are the components of the stress tensor, ε_{kk} is a volumetric strain, ε_{ij} are the components of the strain tensor, λ and μ are material parameters, which are typically referred to as *Lamé coefficients*.

2. Strain-displacement relations

$$\varepsilon = \frac{1}{2}(\nabla \mathbf{u} + \nabla \mathbf{u}^T),$$

where $\mathbf{u} \in \mathbb{R}^3$ is the displacement vector.

3. Equilibrium equation

$$\nabla \cdot \boldsymbol{\sigma} + \rho \mathbf{F} = 0,$$

where $\boldsymbol{\sigma}$ is the stress tensor, $\mathbf{F} \in \mathbb{R}^3$ is the density of external forces, and ρ is the material density.

Thus, the object $\mathbf{Set}_{\mathrm{LE}}$ is formalised via a formalisation functor S into the following model formulation, denoted by $\mathbf{A}_1$:

$$\mathbf{Set}_{\mathrm{LE}} \quad \overset{S}{\mapsto} \quad \begin{cases} \sigma_{ij} = \lambda \delta_{ij} \varepsilon_{kk} + 2\mu \varepsilon_{ij} \\ \varepsilon = \frac{1}{2}(\nabla \mathbf{u} + \nabla \mathbf{u}^T) \\ \nabla \cdot \boldsymbol{\sigma} + \rho \mathbf{F} = 0 \end{cases} \quad =: \mathbf{A}_1.$$

Similarly, the object $\mathbf{Set}_{\mathrm{ME}}$ leads to the model formulation of micropolar elasticity, provided by the following set of equations, see [4] for details:

1. Constitutive equations

$$\begin{aligned} \sigma_{ji} &= (\mu + \alpha)\varepsilon_{ji} + (\mu - \alpha)\varepsilon_{ij} + \lambda \delta_{ij} \varepsilon_{kk}, \\ \mu_{ji} &= (\gamma + \epsilon)\kappa_{ji} + (\gamma - \epsilon)\kappa_{ij} + \beta \delta_{ij} \kappa_{kk}, \end{aligned}$$

where μ_{ji} are the components of the couple stress tensor, σ_{ij} are the components of the force stress tensor, ε_{ij} are the components of the asymmetric tensor of deformation, κ_{ij} are the components of the torsion flexure tensor, $i, j, k = 1, 2, 3$, λ, μ, α, β, γ and ϵ are six micropolar elastic constants.

2. Relations between the components of the strain tensor and displacements and rotations

$$\varepsilon_{ji} = u_{i,j} - e_{kij}\varphi_k, \quad \kappa_{ji} = \varphi_{i,j}, \quad \boldsymbol{\varphi} = \frac{1}{2}\nabla \times \mathbf{u},$$

where u_i is the displacement coordinate, and $\boldsymbol{\varphi} \in \mathbb{R}^3$ is the rotation vector.

3. Equilibrium equations

$$\sigma_{ji,j} + f_i = 0,$$
$$e_{ijk}\sigma_{jk} + \mu_{ji,j} + c_i = 0,$$

where φ_i is the coordinate of the rotation vector, f_i is the coordinate of the body force per unit volume, c_i is the coordinate of the body couple per unit volume, ρ is the density, and e_{ijk} is the permutation symbol.

Thus, the object $\mathbf{Set}_{\mathrm{ME}}$ is formalised via a formalisation functor S into the following model formulation, denoted by $\mathbf{B}_1$:

$$\mathbf{Set}_{\mathrm{ME}} \quad \overset{S}{\mapsto} \quad \begin{cases} \sigma_{ji} = (\mu + \alpha)\varepsilon_{ji} + (\mu - \alpha)\varepsilon_{ij} + \lambda\delta_{ij}\varepsilon_{kk} \\ \mu_{ji} = (\gamma + \epsilon)\kappa_{ji} + (\gamma - \epsilon)\kappa_{ij} + \beta\delta_{ij}\kappa_{kk} \\ \varepsilon_{ji} = u_{i,j} - e_{kij}\varphi_k, \quad \kappa_{ji} = \varphi_{i,j}, \quad \boldsymbol{\varphi} = \dfrac{1}{2}\nabla \times \mathbf{u} \\ \sigma_{ji,j} + f_i = 0 \\ e_{ijk}\sigma_{jk} + \mu_{ji,j} + c_i = 0 \end{cases} =: \mathbf{B}_1.$$

Next, lets us describe the model formulation of nonlinear elasticity associated with the object $\mathbf{Set}_{\mathrm{NE}}$. Moreover, since we consider only the finite strain model—also known as the geometrically nonlinear model—the main difference to the linear model will be visible only in the formulation of the deformation (strain) tensor. There exist various possible formulations of deformation tensors, see for example [18] for details. According to the categorical constructions introduced above, the model formulations obtained by these different tensor formulations are convertible. Thus, we will only present one possible model formulation, which is formalised by the following set of equations:

- Constitutive equations

$$\boldsymbol{\sigma} = \lambda\mathbf{I} + 2\mu\mathbf{C},$$

  where $\boldsymbol{\sigma}$ is the stress tensor, $\mathbf{C}$ is the deformation tensor, λ and μ are Lamé coefficients.
- The right Cauchy-Green deformation tensor

$$\mathbf{C} = \mathbf{F}^T\mathbf{F},$$

  where $\mathbf{F}$ is the deformation gradient tensor.
- Equilibrium equation

$$\nabla \cdot \boldsymbol{\sigma} + \rho\mathbf{K} = 0,$$

  where $\boldsymbol{\sigma}$ is the stress tensor, $\mathbf{K} \in \mathbb{R}^3$ is the density of external forces, and ρ is the material density.

Thus, the object $\mathbf{Set}_{\text{NE}}$ is formalised via a formalisation functor S into the following model formulation, denoted by $\mathbf{C}_1$:

$$\mathbf{Set}_{\text{NE}} \quad \overset{S}{\mapsto} \quad \begin{cases} \boldsymbol{\sigma} = \lambda\mathbf{I} + 2\mu\mathbf{C} \\ \mathbf{C} = \mathbf{F}^T\mathbf{F} \\ \nabla \cdot \boldsymbol{\sigma} + \rho\mathbf{K} = 0 \end{cases} \qquad =: \mathbf{C}_1.$$

Finally, let us discuss the model formulation of linear elastic fracture mechanics derived from the set of assumptions $\mathbf{Set}_{\text{LEF}}$. In fact, the model formulation of linear elastic fracture mechanics is based on the same equations as the linear elasticity model, with the addition of an extra condition specifying the location of cracks:

1. Constitutive equations

$$\sigma_{ij} = \lambda\delta_{ij}\varepsilon_{kk} + 2\mu\varepsilon_{ij}, \quad i, j, k = 1, 2, 3,$$

   where σ_{ij} are the components of the stress tensor, ε_{kk} is a volumetric strain, ε_{ij} are the components of the strain tensor, λ and μ are Lamé coefficients.
2. Strain-displacement relations

$$\varepsilon = \frac{1}{2}(\nabla\mathbf{u} + \nabla\mathbf{u}^T),$$

   where $\mathbf{u} \in \mathbb{R}^3$ is the displacement vector.
3. Equilibrium equation

$$\nabla \cdot \boldsymbol{\sigma} + \rho\mathbf{F} = 0,$$

   where $\boldsymbol{\sigma}$ is the stress tensor, $\mathbf{F} \in \mathbb{R}^3$ is the density of external forces, and ρ is the material density.
4. Geometric information about a crack.

Thus, the object $\mathbf{Set}_{\text{LEF}}$ is formalised via a formalisation functor S into the following model formulation, denoted by $\mathbf{D}_1$:

$$\mathbf{Set}_{\text{LEF}} \quad \overset{S}{\mapsto} \quad \begin{cases} \sigma_{ij} = \lambda\delta_{ij}\varepsilon_{kk} + 2\mu\varepsilon_{ij} \\ \varepsilon = \frac{1}{2}(\nabla\mathbf{u} + \nabla\mathbf{u}^T) \\ \nabla \cdot \boldsymbol{\sigma} + \rho\mathbf{F} = 0 \\ \text{Cracks geometric information} \end{cases} \qquad =: \mathbf{D}_1.$$

It is necessary to make the following remark:

Remark 3.10 Geometric information about the location of cracks is crucial for the model, as the presence of cracks significantly changes the stress field within a body. In particular, a well known crack-tip singularity of the stress field will appear in the linear elastic model, see [6, 15, 16] and references therein. Furthermore, the example of objects $\mathbf{Set}_{\mathrm{LE}}$ and $\mathbf{Set}_{\mathrm{LEF}}$ indicates once more that it is not sufficient to look only at the final model formulations, such as differential equations. As demonstrated by this example, the basic equations remain identical, and the only difference comes from the presence of cracks. Similarly, it is also not sufficient to look only at boundary conditions. For instance, a typical boundary condition for crack faces, which are a part of the boundary, is the traction-free condition implying that normal stresses are zero on crack surfaces. However, zero normal stresses boundary condition on parts of a boundary are a typical condition in many boundary value problems. Therefore, for distinguishing different mathematical models, it is absolutely necessary to analyse the basic modelling assumptions, and not only the final form of equations.

To finish our discussion on the category **Elast**, let us recall that several formulations of elasticity models, namely the Lamé equation and quaternionic formulations, have been mentioned already at the beginning of this chapter. Evidently, similar formulations can be created for the micropolar model, see for example [10, 11], as well as for nonlinear model and linear elastic fracture mechanics model. To keep the presentation short, we will not list various equivalent formulations of elasticity models here, but immediately present the diagram showing the categorical structure at the level of model, see Fig. 3.1.

In fact, the natural transformations shown in Fig. 3.1 might be the same for different models, for example if models are converted to a quaternionic formulations. Additionally, similar to the category **Beam**, formalisation functors S acting on morphisms f_1, f_2, and f_3 might be equal for all considered mathematical models, or might be different if it is desired during the modelling process.

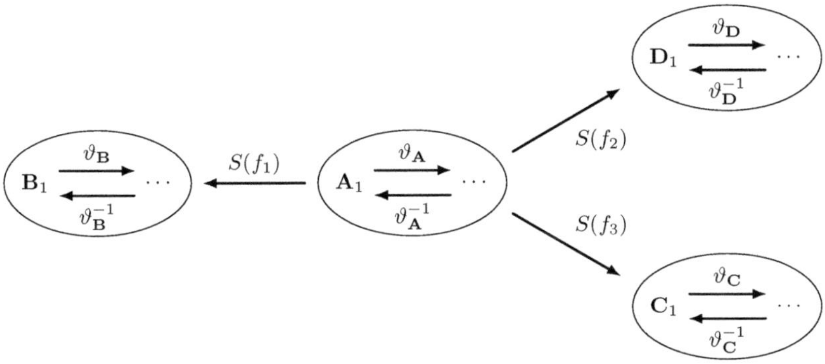

Fig. 3.1 Structure of the category **Elast** at the level of models and where $\vartheta_{\mathbf{i}}$, $i = \{\mathbf{A}, \mathbf{B}, \mathbf{C}, \mathbf{D}\}$, denotes natural transformations appearing in the definition of convertible models, recall Definition 3.4

3.3 Categorical Perspective on Coupled Mathematical Models

In Chap. 1, we have pointed out that modern engineering problems are typically coupled multiphysics problems requiring not a single model, but a combination of several individual models. Consequently, it becomes necessary to extend the categorical perspective on mathematical models presented in previous sections to encompass coupled mathematical models. The idea behind this extension is to preserve the general framework established for categories of individual mathematical models, and elevate it to the level of coupled models. In this way, all constructions introduced for categories of mathematical models in Sect. 3.2 can also be preserved and applied at the level of coupled mathematical models. However, because of a more complicated structure of coupled models, the notions of model complexity and convertibility will also become more technical, as we will see throughout the upcoming sections.

3.3.1 Categories of Coupled Mathematical Models

Similar to the case of categories of "individual" mathematical models, we begin our construction by introducing *concrete* categories $\mathbf{ModCoup}_{i,j}$, $i, j = 1, 2, \ldots$ associated with coupled mathematical models used to describe a certain physical phenomenon, such as models of thermoelasticity. The indices i and j denote the categories of individual mathematical models being coupled. For simplicity, we assume for now that $i \neq j$, and the case $i = j$, i.e. coupling of models from the same category, will be addressed separately later. Moreover, because a coupled mathematical model is still a mathematical model, Definition 3.1 can be straightforwardly extended as follows:

Definition 3.5 Let $\mathbf{Model}_i$ and $\mathbf{Model}_j$ with $i \neq j$ be two categories of mathematical models introduced in Definition 3.1. Then a category $\mathbf{ModCoup}_{i,j}$ is called a *category of coupled mathematical models*, and for all objects of $\mathbf{ModCoup}_{i,j}$ the following assumptions hold:

1. each object is a finite non-empty set—*set of assumptions* of a coupled mathematical model, denoted by $\mathbf{Set}_{\mathrm{A}}$, where A is the corresponding coupled mathematical model;
2. morphisms are relations between these sets;
3. for each set of assumptions and its corresponding model exists a mapping

$$\mathbf{Set}_{\mathrm{A}} \overset{S}{\mapsto} \mathrm{A}.$$

Formally, the above definition describes the structure of categories of coupled mathematical models. As we see from this definition, the assumption on models to act in the same physical dimension has been removed. The reason for removing this assumption comes naturally from the modelling aspects of coupled problems: clearly, coupled problems are not always formulated in the same physical dimension, but rather can combine different domains and different dimensions, as for example in the case of soil-structure interaction or fluid-structure interaction.

Although Definition 3.5 formally describes the structure of a category of coupled mathematical models, the actual structure of this category, especially construction of sets of assumptions, is more intricate. Therefore, we introduce the following refined version of the definition:

Definition 3.6 (Category of Coupled Mathematical Models) Let $\mathbf{Model}_i$ and $\mathbf{Model}_j$ with $i \neq j$ be two categories of mathematical models introduced in Definition 3.1. Then a category $\mathbf{ModCoup}_{i,j}$ is called a *category of coupled mathematical models*, and it is a product category, formally written as

$$\mathbf{ModCoup}_{i,j} = \mathbf{Model}_i \times \mathbf{Model}_j,$$

with projections functors:

$$\mathbf{Model}_i \xleftarrow{\pi_1} \mathbf{Model}_i \times \mathbf{Model}_j \xrightarrow{\pi_2} \mathbf{Model}_j.$$

Moreover, let $\mathbf{Set}_i$ and $\mathbf{Set}_j$ denote objects of categories $\mathbf{Model}_i$ and $\mathbf{Model}_j$, respectively. Then for all objects of $\mathbf{ModCoup}_{i,j}$, denoted by $\mathbf{Set}_{i,j}$, the following assumptions hold:

1. sets of assumptions of coupled mathematical models $\mathbf{Set}_{i,j}$ are constructed formally as follows:

$$\mathbf{Set}_{i,j} := \langle \mathbf{Set}_i \cup \mathbf{Set}_j, \mathbf{Set}_{ij}^{\mathrm{Coup}} \rangle,$$

where the set $\mathbf{Set}_{ij}^{\mathrm{Coup}}$ represents the *coupling conditions*, which are necessary to couple models from $\mathbf{Model}_i$ with models from $\mathbf{Model}_j$. Moreover, the following relations must be satisfied:

$$(a)\ \left(\mathbf{Set}_i \cup \mathbf{Set}_j\right) \subset \mathbf{Set}_{i,j}^{\mathrm{U}}; \quad (b)\ \mathbf{Set}_{i,j}^{\mathrm{U}} \not\subset \left(\mathbf{Set}_i \cup \mathbf{Set}_j\right), \tag{3.5}$$

where $\mathbf{Set}_{i,j}^{\mathrm{U}} = \mathbf{Set}_i \cup \mathbf{Set}_j \cup \mathbf{Set}_{ij}^{\mathrm{Coup}}$;
2. each $\mathbf{Set}_{i,j}$ has the corresponding formalisation mapping S to the respective coupled mathematical model.

Let us discuss assumption 1 from the last definition in more details. As it is stated in the definition, the set of assumptions of a coupled mathematical model is not just a union of the sets of assumptions of individual models $\mathbf{Set}_i \cup \mathbf{Set}_j$, but a union of these individual sets of assumptions with a new set $\mathbf{Set}_{ij}^{\text{Coup}}$, which contains coupling conditions arising during the model coupling process.

It is also important to mention that originally in works [9, 13] the definition of a category of coupled mathematical models has been formulated a bit differently. Particularly, sets of assumptions of coupled mathematical models have been originally constructed as follows:

$$\mathbf{Set}_{i,j} := \mathbf{T}\left(\mathbf{Set}_i\right) \cup \mathbf{F}\left(\mathbf{Set}_j\right),$$

where $\mathbf{T}$ and $\mathbf{F}$ are functorial mappings between $\mathbf{Model}_i$, $\mathbf{Model}_j$ and $\mathbf{ModCoup}_{i,j}$, respectively. The idea was to underline the fact that a coupling of models is more than a simple union of these models, and, therefore, during the coupling process sets of assumptions of individual models are modified by functors acting on them. However, from the modelling perspective, the sets of assumptions of individual models are not really modified, but rather extra coupling conditions are added to both sets. Nonetheless, the construction of sets of assumptions of coupled models by using functors does not exclude this situation, because modification of objects $\mathbf{Set}_i$, $\mathbf{Set}_j$ by functors implies adding something more to the both objects. However, for the sake of clarity, the fact that coupling is more than just a union has been stated explicitly in Definition 3.6 in the construction of objects $\mathbf{Set}_{i,j}$.

Another essential difference to the definition presented in [9, 13] is that according to Definition 3.6, objects in categories of coupled mathematical models are now pairs. This structural difference will help us to discuss complexity of coupled mathematical models in a more comprehensive way.

Let us now make the following remark:

Remark 3.11 For practical applications and interpretations, it might be beneficial to view the projection functors π_1 and π_2 appearing in Definition 3.6 as *decoupling* process for coupled models.

As a summary of the above discussion, we can now elaborate how coupled model are constructed in practice. This construction will be done again by the help of functorial mappings, as it has been done in [9, 13], but the role of these functors will be clarified from the modelling perspective. We introduce the following proposition:

Proposition 3.2 *Let* $\mathbf{Model}_i$ *and* $\mathbf{Model}_j$ *with* $i \neq j$ *be two categories of mathematical models, and let* $\mathbf{Set}_i$ *and* $\mathbf{Set}_j$ *denote objects of categories* $\mathbf{Model}_i$ *and* $\mathbf{Model}_j$*, respectively. The coupling of categories* $\mathbf{Model}_i$ *and* $\mathbf{Model}_j$ *is realised by help of functorial mappings acting on objects* $\mathbf{Set}_i$ *and* $\mathbf{Set}_j$ *as follows:*

$$\mathbf{T}\left(\mathbf{Set}_i\right) = \langle \mathbf{Set}_i, \mathbf{Set}_i^{\text{Coup}} \rangle, \quad \mathbf{F}\left(\mathbf{Set}_j\right) = \langle \mathbf{Set}_j, \mathbf{Set}_j^{\text{Coup}} \rangle,$$

and, consequently, the objects $\mathbf{Set}_{i,j}$ *of category* $\mathbf{ModCoup}_{i,j}$ *are constructed formally as follows:*

$$\begin{aligned}\mathbf{Set}_{i,j} &:= \mathbf{T}(\mathbf{Set}_i) \cup \mathbf{F}\left(\mathbf{Set}_j\right) \\ &= \langle \mathbf{Set}_i \cup \mathbf{Set}_j, \mathbf{Set}_i^{\mathrm{Coup}} \cup \mathbf{Set}_j^{\mathrm{Coup}} \rangle \\ &= \langle \mathbf{Set}_i \cup \mathbf{Set}_j, \mathbf{Set}_{ij}^{\mathrm{Coup}} \rangle .\end{aligned}$$

Then for $\mathbf{Set}_i^{\mathrm{Coup}}$ *and* $\mathbf{Set}_j^{\mathrm{Coup}}$ *one of the following statements holds:*

1. *both sets* $\mathbf{Set}_i^{\mathrm{Coup}}$ *and* $\mathbf{Set}_j^{\mathrm{Coup}}$ *are not empty;*
2. *only one of the sets* $\mathbf{Set}_i^{\mathrm{Coup}}$ *and* $\mathbf{Set}_j^{\mathrm{Coup}}$ *is not empty.*

Proof The proof of this proposition will be done by the help of modelling arguments, and it is based directly on Definition 3.6. The first statement can be straightforwardly proved by assuming the contrary—both sets $\mathbf{Set}_i^{\mathrm{Coup}}$ and $\mathbf{Set}_j^{\mathrm{Coup}}$ are empty sets. This implies that the relations (3.5) are not satisfied, because the relation b) does not hold in the case $\mathbf{Set}_{ij}^{\mathrm{Coup}} = \emptyset$. Thus, both sets cannot be empty simultaneously. Practically it means again, that a coupling of models is not just a union of two models (the case corresponding to both sets $\mathbf{Set}_i^{\mathrm{Coup}}$ and $\mathbf{Set}_j^{\mathrm{Coup}}$ being empty), but it requires formulating extra coupling conditions to satisfy the relations (3.5).

The proof of the second case is also straightforward, and it follows immediately from Definition 3.6. Obviously that the case when one of the sets $\mathbf{Set}_i^{\mathrm{Coup}}$ and $\mathbf{Set}_j^{\mathrm{Coup}}$ is empty satisfies the relations (3.5), and thus this case represents valid objects in categories of coupled mathematical models. Practically, it means that coupling is realised only from one side, i.e. only one sub-model is influenced by another model. □

Before introducing the notion of complexity at the level of coupled models, it is important to also consider the case of coupling of models from the same category, i.e. when $i = j$. This case has been omitted in the discussion above, but it can also appear in practical applications. A typical example illustrating the need for such a coupling is the heat conduction problem formulated in a domain composed of two different materials:

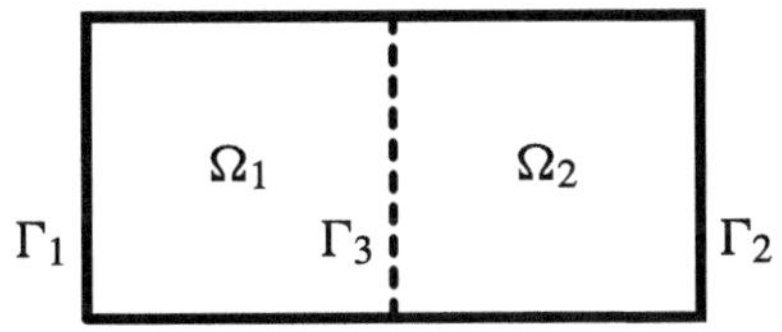

Here domain Ω_1 has material parameters c_1, ρ_1, λ_1, while domain Ω_2 has material parameters c_2, ρ_2, λ_2, and Γ_3 is the coupling interface where coupling conditions for the heat flux and temperature field must be formulated.

To address the heat conduction problem in $\Omega = \Omega_1 \cup \Omega_2$, we can formally consider two categories $\mathbf{Heat}_1$ and $\mathbf{Heat}_2$ of mathematical models for heat conduction in Ω_1 and Ω_2, respectively. In this case, Definition 3.6 can be used directly to describe the coupled heat conduction models for Ω. The drawback of this approach, however, is the fact that categories $\mathbf{Heat}_1$ and $\mathbf{Heat}_2$ contain exactly the same objects, which leads to unnecessary duplication of categories. Furthermore, this problem will become even more confusing if more than two models need to be coupled. Therefore, for the sake of clarity, it is beneficial to propose an alternative approach to couple models, which formally must belong to the same category.

To address the coupling of models from the same category, we propose to work not with a product of categories, but with its dual object—*coproduct*. Consequently, we present the following definition:

Definition 3.7 Let $\mathbf{Model}_i$ be a category of mathematical models introduced in Definition 3.1. Then a category $\mathbf{ModCoup}_{i,i}$ is called a *category of coupled mathematical models from the same category*, and it is a coproduct, formally written as

$$\mathbf{ModCoup}_{i,i} = \mathbf{Model}_i + \mathbf{Model}_i,$$

with injections:

$$\mathbf{Model}_i \xrightarrow{i_1} \mathbf{Model}_i + \mathbf{Model}_i \xleftarrow{i_2} \mathbf{Model}_i.$$

Moreover, for all objects of $\mathbf{ModCoup}_{i,i}$, denoted by $\mathbf{Set}_{i,i}$, the following assumptions hold:

1. sets of assumptions of a coupled mathematical models $\mathbf{Set}_{i,i}$ are constructed formally as follows:

$$\mathbf{Set}_{i,i} := \langle \mathbf{Set}_i + \mathbf{Set}_i, \mathbf{Set}_{ii}^{\text{Coup}} \rangle,$$

   where the set $\mathbf{Set}_{ii}^{\text{Coup}} \neq \emptyset$ represents the *coupling conditions*, which are necessary to couple models from $\mathbf{Model}_i$;
2. each $\mathbf{Set}_{i,i}$ has the corresponding formalisation mapping S to the respective coupled mathematical model.

The use of a disjoint union in construction of $\mathbf{Set}_{i,i}$ helps to avoid the situation when coupling of models from the same category leads to a nonsensical result, as it could be in the case of using a standard union. The nonsensical result appears for example when we try to couple models in the sense of Definition 3.6, which are complexity-related. Because models are complexity-related, the intersection of their corresponding sets of assumptions is not empty. Therefore, the union of these sets would contain only basic modelling assumptions corresponding to a simpler model. Thus, the coupled model in this case would be a pair containing basic modelling assumptions of the simplest model and a set of coupling conditions. Obviously, this result is nonsensical. This problem can be avoided by using a disjoint union, as proposed in Definition 3.7, because in this case all assumptions of individual models are preserved and only the set of coupling conditions is added.

A more detailed discussion on construction of objects $\mathbf{Set}_{i,i}$ is provided by the following corollary, which is a simplified version of Proposition 3.2:

Corollary 3.3 *Let* $\mathbf{Model}_i$ *be a category of mathematical models with objects denoted by* $\mathbf{Set}_i$*. The coupling of category* $\mathbf{Model}_i$ *with itself is realised by help of functorial mappings acting on objects* $\mathbf{Set}_i$ *as follows:*

$$\mathbf{T}\left(\mathbf{Set}_i\right) = \langle \mathbf{Set}_i, \mathbf{Set}_{\mathbf{T}}^{\mathrm{Coup}} \rangle, \quad \mathbf{F}\left(\mathbf{Set}_j\right) = \langle \mathbf{Set}_j, \mathbf{Set}_{\mathbf{F}}^{\mathrm{Coup}} \rangle,$$

and, consequently, the objects $\mathbf{Set}_{i,i}$ *of category* $\mathbf{ModCoup}_{i,i}$ *are constructed formally as follows:*

$$\begin{aligned} \mathbf{Set}_{i,i} &:= \mathbf{T}\left(\mathbf{Set}_i\right) \cup \mathbf{F}\left(\mathbf{Set}_i\right) \\ &= \langle \mathbf{Set}_i + \mathbf{Set}_j, \mathbf{Set}_{\mathbf{T}}^{\mathrm{Coup}} \cup \mathbf{Set}_{\mathbf{F}}^{\mathrm{Coup}} \rangle \\ &= \langle \mathbf{Set}_i + \mathbf{Set}_j, \mathbf{Set}_{ii}^{\mathrm{Coup}} \rangle. \end{aligned}$$

It is necessary to make the following remark:

Remark 3.12 The objects $\mathbf{Set}_{\mathbf{T}}^{\mathrm{Coup}}$ and $\mathbf{Set}_{\mathbf{F}}^{\mathrm{Coup}}$ in general will contain the same set of coupling conditions, because coupling of models from the same category is considered. However, for the sake of generality, we treat these sets as unique objects here.

Finally, let us briefly discuss how Definitions 3.6–3.7 are reflected at the level of models. In this context, Definition 3.6 corresponds to the following commutative diagram:

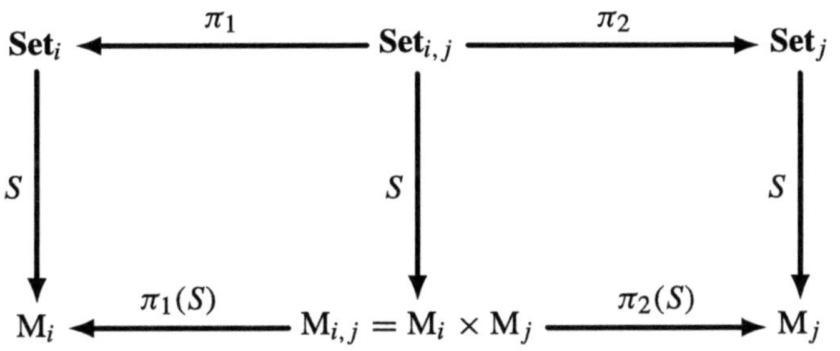

Here M_i, M_j, and $\mathrm{M}_{i,j}$ are instances (model formulations) of the corresponding sets of assumptions. Moreover, the formal product $\mathrm{M}_i \times \mathrm{M}_j$ indicates that models representing different physical phenomena, i.e. belonging to different categories, have been coupled. A similar commutative diagram can be drawn for Definition 3.7:

$$
\begin{array}{ccccc}
\mathbf{Set}_i & \xrightarrow{\;i_1\;} & \mathbf{Set}_{i,i} & \xleftarrow{\;i_2\;} & \mathbf{Set}_i \\
\Big\downarrow S & & \Big\downarrow S & & \Big\downarrow S \\
\mathrm{M}_i & \xrightarrow{\;i_1(S)\;} & \mathrm{M}_{i,i} = M_i + M_i & \xleftarrow{\;i_2(S)\;} & \mathrm{M}_i
\end{array}
$$

where we have taken into account the coproduct structure. The formal disjoint union $M_i + M_i$ indicates that the models describing the same physical phenomenon, i.e. belonging to the same category, have been coupled.

3.3.2 *Relations Between Coupled Mathematical Models*

After introducing the basic structure of categories of coupled mathematical models in terms of product and coproduct, it is necessary to address the question of relations between different coupled models. Hence, the notion of *model complexity* must be generalised to coupled models. However, this generalisation is not so straightforward due to the structure of categories of coupled models, in particular because of the sets of coupling conditions $\mathbf{Set}_{ij}^{\mathrm{Coup}}$ and $\mathbf{Set}_{ii}^{\mathrm{Coup}}$ for coupling models from different categories and from the same category, respectively. As a result, we need to distinguish the complexity of coupled models on two levels:

- Complexity with respect to the model assumptions of models being coupled, which will be called *base complexity*.
- Complexity with respect to the sets of coupling conditions, which will be called *coupling complexity*.

To address both levels, the following definition of model complexity for coupled models is proposed:

Definition 3.8 (Complexity of Coupled Mathematical Models) Let $\mathbf{Model}_i$ and $\mathbf{Model}_j$ be two categories of mathematical models, and let $\mathbf{ModCoup}_{i,j}$ be the category of coupled mathematical models. Further, let A and B be mathematical models in the category $\mathbf{ModCoup}_{i,j}$ with sets of assumptions $\mathbf{Set}_{\mathrm{A}}$, $\mathbf{Set}_{\mathrm{B}}$, and sets of coupling conditions $\mathbf{Set}_{\mathrm{A}}^{\mathrm{Coup}}$, $\mathbf{Set}_{\mathrm{B}}^{\mathrm{Coup}}$, respectively. Then

- the model A has *higher base complexity* than the model B if and only if $\mathbf{Set}_{\mathrm{A}} \subset \mathbf{Set}_{\mathrm{B}}$, but $\mathbf{Set}_{\mathrm{B}} \not\subset \mathbf{Set}_{\mathrm{A}}$;
- the model A has *higher coupling complexity* than the model B if and only if $\mathbf{Set}_{\mathrm{A}}^{\mathrm{Coup}} \subset \mathbf{Set}_{\mathrm{B}}^{\mathrm{Coup}}$, but $\mathbf{Set}_{\mathrm{B}}^{\mathrm{Coup}} \not\subset \mathbf{Set}_{\mathrm{A}}^{\mathrm{Coup}}$;

- the model A has *higher complexity* than model B if and only if it is simultaneously has higher base and coupling complexity.

Consequently, three levels of equality of coupled models in the sense of complexity can be distinguished:

- the models A and B are called *equal in base* if and only if $\mathbf{Set}_{\mathrm{B}} = \mathbf{Set}_{\mathrm{A}}$;
- the models A and B are called *equal in coupling* if and only if $\mathbf{Set}_{\mathrm{A}}^{\mathrm{Coup}} = \mathbf{Set}_{\mathrm{B}}^{\mathrm{Coup}}$;
- the models A and B are called *equal* if and only if they are equal in base and in coupling.

It is important to remark that similar to the model complexity introduced in Definition 3.2, the complexity of a coupled model represents a relative quality measure of how good the coupled model represents the coupled problem under consideration. Furthermore, the proposed definition of complexity for coupled models reflects the fact, that coupled problems involve different models being coupled, which can be of different complexity, as well as the coupling itself can be realised with different levels of detail. These facts are captured by the notion of base complexity and coupling complexity.

Evidently, Definition 3.8 is applicable for the case when models being coupled belong to different categories, as well as for the case when models being coupled belong to the same category. Additionally, in some practical situations it might be beneficial to explicitly underline in the case of base complexity which individual model provides a more complex contribution to the overall complexity of a coupled model. To address this situation, we refine the notion of base complexity as follows:

Definition 3.9 (Refinement of Base Complexity) Let $\mathbf{Model}_i$ and $\mathbf{Model}_j$ be two categories of mathematical models, and let $\mathbf{ModCoup}_{i,j}$ be the category of coupled mathematical models. Further, let A and B be mathematical models in the category $\mathbf{ModCoup}_{i,j}$, which are related by help of base complexity. Then the following two cases can be distinguished:

- the model A has *higher left-base complexity* than the model B if and only if

$$\pi_1\left(\mathbf{Set}_{\mathrm{A}}\right) \subset \pi_1\left(\mathbf{Set}_{\mathrm{B}}\right), \text{ but } \pi_1\left(\mathbf{Set}_{\mathrm{A}}\right) \not\subset \pi_1\left(\mathbf{Set}_{\mathrm{B}}\right) \text{ for } i \neq j$$

  or

$$i_1\left(\mathbf{Set}_{\mathrm{A}}\right) \subset i_1\left(\mathbf{Set}_{\mathrm{B}}\right), \text{ but } i_1\left(\mathbf{Set}_{\mathrm{A}}\right) \not\subset i_1\left(\mathbf{Set}_{\mathrm{B}}\right) \text{ for } i = j;$$

- the model A has *higher right-base complexity* than the model B if and only if

$$\pi_2\left(\mathbf{Set}_{\mathrm{A}}\right) \subset \pi_2\left(\mathbf{Set}_{\mathrm{B}}\right), \text{ but } \pi_2\left(\mathbf{Set}_{\mathrm{A}}\right) \not\subset \pi_2\left(\mathbf{Set}_{\mathrm{B}}\right) \text{ for } i \neq j,$$

  or

$$i_2\left(\mathbf{Set}_{\mathrm{A}}\right) \subset i_2\left(\mathbf{Set}_{\mathrm{B}}\right), \text{ but } i_2\left(\mathbf{Set}_{\mathrm{A}}\right) \not\subset i_2\left(\mathbf{Set}_{\mathrm{B}}\right) \text{ for } i = j.$$

The next proposition is a direct consequence of the last definition and our discussion on partially and totally ordered categories in Sect. 3.2:

Proposition 3.3 *Let* $\mathbf{ModCoup}_{i,j}$ *be a category of coupled mathematical models in which n objects are related by the help of base complexity. Then the following statements hold about categories of mathematical models* $\mathbf{Model}_i$ *and* $\mathbf{Model}_j$*:*

1. *if n objects of* $\mathbf{ModCoup}_{i,j}$ *are left-base complexity related, then* $\mathbf{Model}_i$ *contains a totally ordered subcategory;*
2. *if n objects of* $\mathbf{ModCoup}_{i,j}$ *are right-base complexity related, then* $\mathbf{Model}_j$ *contains a totally ordered subcategory;*
3. *if n objects of* $\mathbf{ModCoup}_{i,j}$ *are left- and right-base complexity related, then* $\mathbf{Model}_i$ *and* $\mathbf{Model}_j$ *contain totally ordered subcategories.*

Proof The proof of this proposition is straightforward, and it is based on the use of Definition 3.6 and the notion of totally and partially ordered categories. Let us illustrate the idea of the proof on the case 1. By using the definition of categories of coupled mathematical models and the fact that n objects of $\mathbf{ModCoup}_{i,j}$ are left-base complexity related, we obtain the following (left-side) diagram

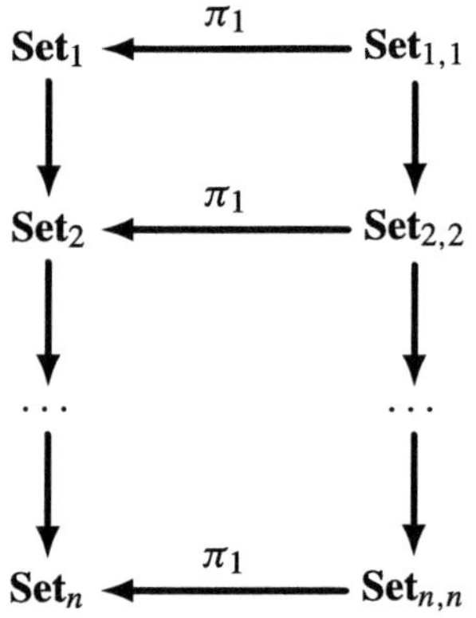

Thus, after using first projection π_1 we obtain n objects in category $\mathbf{Model}_i$ which are complexity related in the sense of Definition 3.2, and, therefore, constitute a totally ordered subcategory with n objects. The cases 2 and 3 are obtained analogously by considering the right-side and full diagram, respectively. The case when mathematical models from the same category are coupled can be proved analogously and instead of working with projections π_1 and π_2, inclusions i_1 and i_2 must be considered. □

It is necessary to make the following remark regarding the last proposition:

Remark 3.13 We have used “subcategory” in Proposition 3.3, because only n objects of $\mathbf{Model}_i$ and $\mathbf{Model}_j$ actually used in construction of the category $\mathbf{ModCoup}_{i,j}$ of coupled mathematical models are considered. In general, categories $\mathbf{Model}_i$ and $\mathbf{Model}_j$ can be partially ordered, if they contain more models, which are not complexity related and are not used in $\mathbf{ModCoup}_{i,j}$, or can be totally ordered if they contain exactly n objects used for constructing the category $\mathbf{ModCoup}_{i,j}$.

The results related to the totally and partially ordered categories of mathematical models, as presented in Sect. 3.2, can be directly extended to the case of coupled models. The main difference is that three notions of complexity are introduced for coupled models, allowing total and partial orders to be defined with respect to these complexity. Because the reasoning and proofs for the results on totally and partially ordered categories of coupled mathematical models closely follow those provided in Sect. 3.2, we will keep the presentation concise, offering only a few comments and remarks on the way.

The following definition introduces totally and partially ordered categories of coupled mathematical models:

Definition 3.10 Let $\mathbf{Model}_i$ and $\mathbf{Model}_j$ be two categories of mathematical models, and let $\mathbf{ModCoup}_{i,j}$ be the category of coupled mathematical models. Let $\mathbf{Model}_i$ contains n objects $\mathbf{Set}_{A_k}$, $k = 1, \ldots, n$, and let $\mathbf{Model}_j$ contains m objects $\mathbf{Set}_{B_s}$, $s = 1, \ldots, m$. Further, let $\mathbf{Set}_l^{\text{Coup}}$, $l = 1, \ldots, N$ be sets of coupling conditions which can be ordered according to Definition 3.8. Moreover, let X be the set of all coupling conditions used in the category $\mathbf{ModCoup}_{i,j}$. Then the following cases can be distinguished:

- if the category $\mathbf{Model}_i$ is a totally ordered category according to Definition 3.3, then the category $\mathbf{ModCoup}_{i,j}$ is called *totally ordered on the left*;
- if the category $\mathbf{Model}_j$ is a totally ordered category according to Definition 3.3, then the category $\mathbf{ModCoup}_{i,j}$ is called *totally ordered on the right*;
- the category $\mathbf{ModCoup}_{i,j}$ is called *totally ordered in base*, if it is totally ordered on the left and on the right;
- the category $\mathbf{ModCoup}_{i,j}$ is called *totally ordered in coupling*, if and only if

$$X = \mathbf{Set}_1^{\text{Coup}} \cup \mathbf{Set}_2^{\text{Coup}} \cup \ldots \cup \mathbf{Set}_N^{\text{Coup}}, \text{ and } \mathbf{Set}_N^{\text{Coup}} = X;$$

- finally, the category $\mathbf{ModCoup}_{i,j}$ is called *totally ordered* if it is totally ordered in base and in coupling.

If non of the above cases is applicable, then the category $\mathbf{ModCoup}_{i,j}$ is simply called partially ordered.

It is worth noting, that this definition allows considering different number of objects in categories $\mathbf{Model}_i$ and $\mathbf{Model}_j$ being coupled. The meaning of this evidently comes from the fact that different categories of mathematical models might have a different number of objects and nonetheless can be coupled in various combinations into coupled models. This point will become even clearer during our discussion on constructing comma categories in the next section.

Next, we extend the notion of the most complex and the simplest objects to coupled models. Again, it is necessary to take into account different versions of total ordering in categories of coupled mathematical models. We have now the following corollary:

Corollary 3.4 *Let* $\mathbf{ModCoup}_{i,j}$ *be the category of coupled mathematical models for which one of total ordering introduced in Definition 3.10 holds, then*

- *if the category* $\mathbf{ModCoup}_{i,j}$ *is totally ordered on the left, then the category* $\mathbf{Model}_i$ *contains the most complex and the simplest objects, which are unique;*
- *if the category* $\mathbf{ModCoup}_{i,j}$ *is totally ordered on the right, then the category* $\mathbf{Model}_j$ *contains the most complex and the simplest objects, which are unique;*
- *if the category* $\mathbf{ModCoup}_{i,j}$ *is totally ordered in base, then it contains the most complex and the simplest objects w.r.t. base, which are unique;*
- *if the category* $\mathbf{ModCoup}_{i,j}$ *is totally ordered in coupling, then it contains the most complex and the simplest objects w.r.t. coupling, which are unique.*

Proposition 3.1 also straightforwardly extends to the case of coupled models:

Proposition 3.4 *Let* $\mathbf{ModCoup}_{i,j}$ *be a partially ordered category with N objects, and let* $\mathbf{Model}_i$ *contains n objects* $\mathbf{Set}_{A_k}$, $k = 1, \ldots, n$, *and* $\mathbf{Model}_j$ *contains m objects* $\mathbf{Set}_{B_s}$, $s = 1, \ldots, m$. *Then the following statements can be made:*

1. *For the categories* $\mathbf{Model}_i$ *and* $\mathbf{Model}_j$ *one the following cases holds:*
   (a) *the most complex object* $\mathbf{Set}_{\mathrm{A}_1}$ *(or* $\mathbf{Set}_{B_1}$*) and the simplest object* $\mathbf{Set}_{\mathrm{A}_n}$ *(or* $\mathbf{Set}_{B_m}$*) do not exist;*
   (b) *the most complex object* $\mathbf{Set}_{\mathrm{A}_1}$ *(or* $\mathbf{Set}_{B_1}$*) exists, while the simplest object* $\mathbf{Set}_{\mathrm{A}_n}$ *(or* $\mathbf{Set}_{B_m}$*) does not exist;*
   (c) *the most complex object* $\mathbf{Set}_{\mathrm{A}_1}$ *(or* $\mathbf{Set}_{B_1}$*) does not exist, while the simplest object* $\mathbf{Set}_{\mathrm{A}_n}$ *(or* $\mathbf{Set}_{B_m}$*) exists;*
   (d) *the most complex object* $\mathbf{Set}_{\mathrm{A}_1}$ *(or* $\mathbf{Set}_{B_1}$*) and the simplest object* $\mathbf{Set}_{\mathrm{A}_n}$ *(or* $\mathbf{Set}_{B_m}$*) exist simultaneously.*
2. *If the category* $\mathbf{ModCoup}_{i,j}$ *is partially ordered in base, then one of the following cases holds:*
   (a) *the most complex object* $\mathbf{Set}_1$ *and the simplest object* $\mathbf{Set}_N$ *w.r.t. base do not exist;*
   (b) *the most complex object* $\mathbf{Set}_1$ *w.r.t. base exists, while the simplest object* $\mathbf{Set}_N$ *w.r.t. base does not exist;*
   (c) *the most complex object* $\mathbf{Set}_1$ *w.r.t. base does not exist, while the simplest object* $\mathbf{Set}_N$ *w.r.t. base exists;*
   (d) *the most complex object* $\mathbf{Set}_1$ *w.r.t. base and the simplest object* $\mathbf{Set}_N$ *w.r.t. base exist simultaneously.*
3. *If the category* $\mathbf{ModCoup}_{i,j}$ *is partially ordered in coupling, then one of the following cases holds:*
   (a) *the most complex object* $\mathbf{Set}_1^{\mathrm{Coup}}$ *and the simplest object* $\mathbf{Set}_N^{\mathrm{Coup}}$ *w.r.t. coupling do not exist;*

(b) *the most complex object* $\mathbf{Set}_1^{\mathrm{Coup}}$ *w.r.t. coupling exists, while the simplest object* $\mathbf{Set}_N^{\mathrm{Coup}}$ *w.r.t. coupling does not exist;*
(c) *the most complex object* $\mathbf{Set}_1^{\mathrm{Coup}}$ *w.r.t. coupling does not exist, while the simplest object* $\mathbf{Set}_N^{\mathrm{Coup}}$ *w.r.t. coupling exists;*
(d) *the most complex object* $\mathbf{Set}_1^{\mathrm{Coup}}$ *w.r.t. coupling and the simplest object* $\mathbf{Set}_N^{\mathrm{Coup}}$ *w.r.t. coupling exist simultaneously.*

4. *Finally, if the category* $\mathbf{ModCoup}_{i,j}$ *is partially ordered in base and in coupling, then combinations of cases under 2 and 3 are possible.*

Proof The proof of 1 is, in fact, follows immediately from Proposition 3.1. Moreover, to prove 2 and 3 we need to apply the same constructive proof as for Proposition 3.1. The last case 4 is then follows directly from 2 and 3 by considering various combinations of the respective cases and by taking into account Definition 3.6 and construction of objects in the category $\mathbf{ModCoup}_{i,j}$. From this construction, it follows that sets $\mathbf{Set}_i \cup \mathbf{Set}_j$ and $\mathbf{Set}_{ij}^{\mathrm{Coup}}$ are considered independently, and hence, various combinations of cases 2 and 3 can be constructed without any extra conditions. Thus, the proposition is proved. □

Next, we have the following theorem:

Theorem 3.3 *Consider a category* $\mathbf{ModCoup}_{i,j}$ *with N objects, and let* $\mathbf{Model}_i$ *contains n objects, and* $\mathbf{Model}_j$ *contains m objects. Then the following cases hold:*

- *If the most complex object and the simplest object exist simultaneously in the category* $\mathbf{Model}_i$ *(*$\mathbf{Model}_j$*), then* $\mathbf{Model}_i$ *(*$\mathbf{Model}_j$*) is a totally ordered category;*
- *If the most complex object and the simplest object w.r.t. base exist simultaneously in the category* $\mathbf{ModCoup}_{i,j}$*, then* $\mathbf{ModCoup}_{i,j}$ *is a totally ordered in base;*
- *If the most complex object and the simplest object w.r.t. coupling exist simultaneously in the category* $\mathbf{ModCoup}_{i,j}$*, then* $\mathbf{ModCoup}_{i,j}$ *is a totally ordered in coupling;*
- *If the most complex object and the simplest object w.r.t. base and w.r.t. coupling exist simultaneously in the category* $\mathbf{ModCoup}_{i,j}$*, then* $\mathbf{ModCoup}_{i,j}$ *is a totally ordered category.*

Proof The proof of the theorem is identical to the proof of Theorem 3.1, requiring only an individual consideration of each case and a proof by contradiction. □

Finally, the above discussion on connection between partially and totally ordered categories of coupled models can be summarised by the following theorem:

Theorem 3.4 *Every partially ordered category of coupled mathematical models contains at least one totally ordered category of coupled mathematical models as a subcategory: either w.r.t. base, or w.r.t. coupling, or both.*

3.3.3 Convertibility of Coupled Mathematical Models

The next step is to extend the notion of convertible mathematical models introduced in Definition 3.4 to coupled mathematical models. Intuitively, it is clear how to define convertible coupled mathematical models. In fact, it is necessary to take into account, that functorial mappings must act also on the set of coupling conditions $\mathbf{Set}_{ij}^{\text{Coup}}$, as indicated by the diagram:

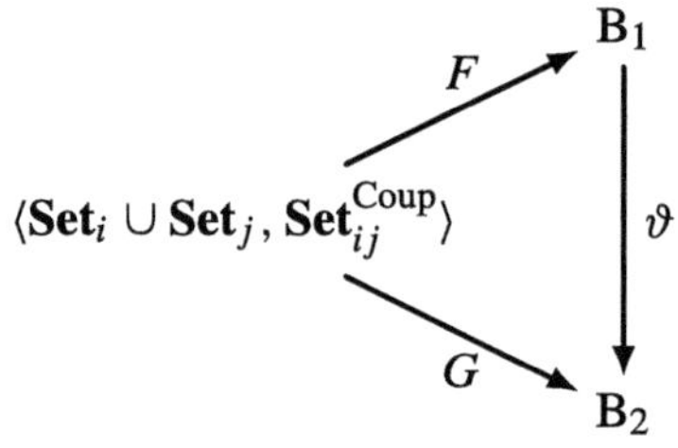

with

$$\vartheta : F\left(\langle \mathbf{Set}_i \cup \mathbf{Set}_j, \mathbf{Set}_{ij}^{\text{Coup}} \rangle\right) \to G\left(\langle \mathbf{Set}_i \cup \mathbf{Set}_j, \mathbf{Set}_{ij}^{\text{Coup}} \rangle\right).$$

However, to build on the discussion from the previous subsection, a more detailed analysis of convertible coupled mathematical models is required. Nonetheless, we start with the following definition, which is a direct extension of Definition 3.4:

Definition 3.11 Let $\mathbf{Set}_\text{A}$ be an object in the category **ModCoup**, and let B_1 and B_2 be two possible formulations of the coupled model associated with the object $\mathbf{Set}_\text{A}$ via two functors F and G. Then the model formulations B_1 and B_2 are connected via a natural transformation of functors ϑ, and the model formulations B_1 and B_2 are called *convertible*. This construction corresponds to the commutative diagram

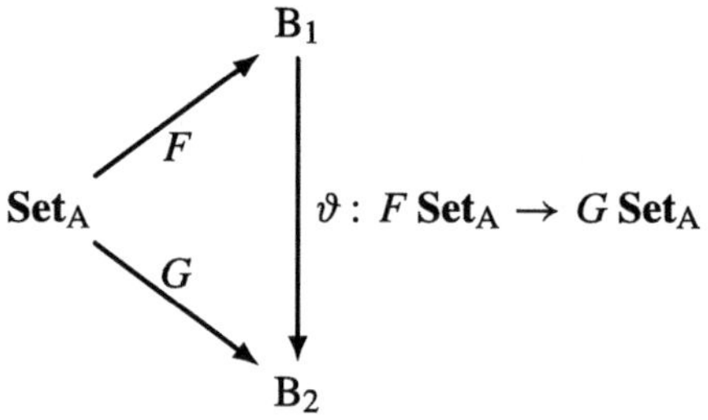

Moreover, models which are instantiated by convertible model formulations will be called *convertible models*.

This definition represents a very general point of view on the convertibility of coupled mathematical models, or, in fact, of any mathematical model. Strictly speaking, this point of view neglects the way how coupled models are created, meaning that only the final set of assumptions of a coupled mathematical model and its formalisation mapping are considered. However, considering Definitions 3.6–3.7 and discussion around them, it is beneficial to make Definition 3.11 more precise and distinguish between convertible models on the left, on the right, and in coupling. Consequently, we introduce now the following definition:

Definition 3.12 Let $\mathbf{Model}_i$ and $\mathbf{Model}_j$ be two categories of mathematical models, and let $\mathbf{ModCoup}_{i,j}$ be the category of coupled mathematical models. Let $\mathbf{Set}_{i,j}$ be an object in the category $\mathbf{ModCoup}_{i,j}$, and let B_1 and B_2 be two possible formulations of the coupled model associated with the object $\mathbf{Set}_{i,j}$ via two functors F and G. Then the following notions of convertibility can be distinguished:

- the model formulations B_1 and B_2 are called *convertible on the left*, if the diagram commutes

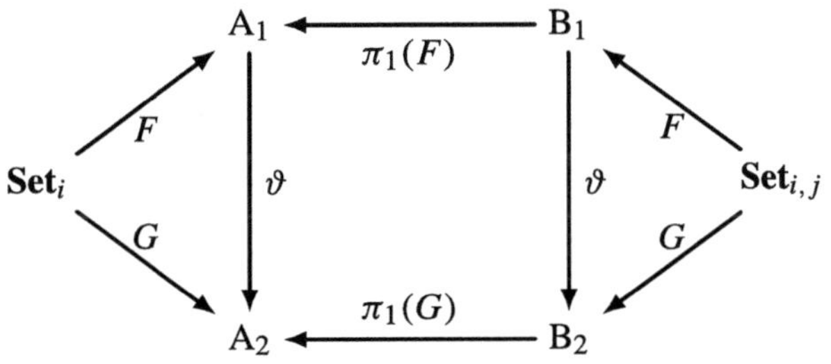

for $i \neq j$, and where $\vartheta : F\,\mathbf{Set}_{i,j} \to G\,\mathbf{Set}_{i,j}$ is a natural transformation, or more precisely, $\vartheta : F\,\mathbf{Set}_i \to G\,\mathbf{Set}_i$. Similarly for $i = j$, the following diagram must commute

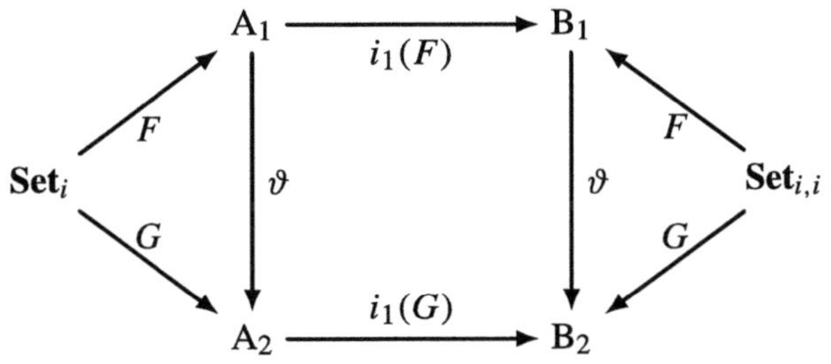

- the model formulations B_1 and B_2 are called *convertible on the right*, if the diagram commutes

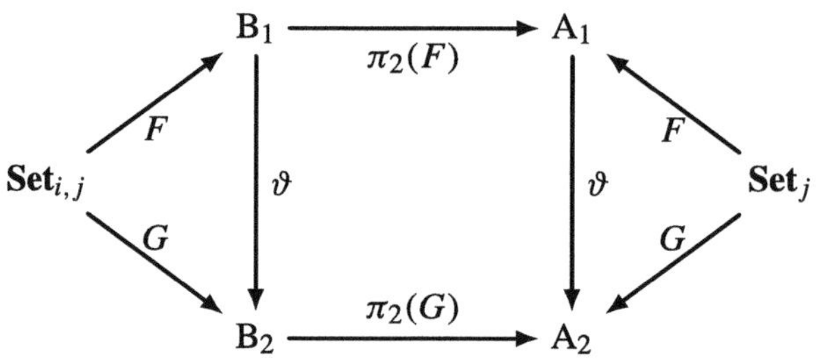

for $i \neq j$, and where $\vartheta : F\,\mathbf{Set}_{i,j} \to G\,\mathbf{Set}_{i,j}$ is a natural transformation, or more precisely, $\vartheta : F\,\mathbf{Set}_j \to G\,\mathbf{Set}_j$. Similarly for $i = j$, the following diagram must commute

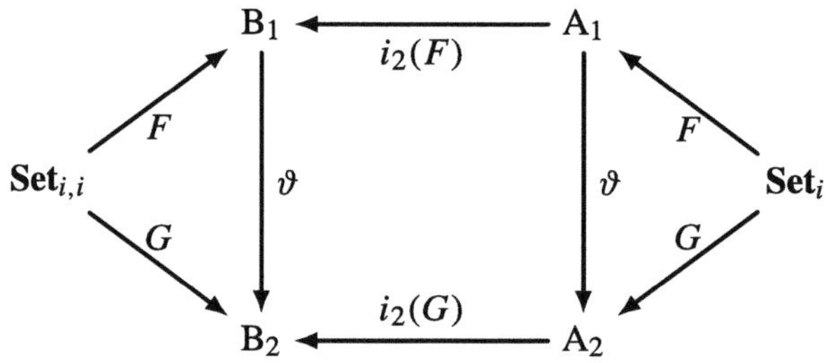

- the model formulations B_1 and B_2 are called *convertible in coupling*, if the diagrams commute

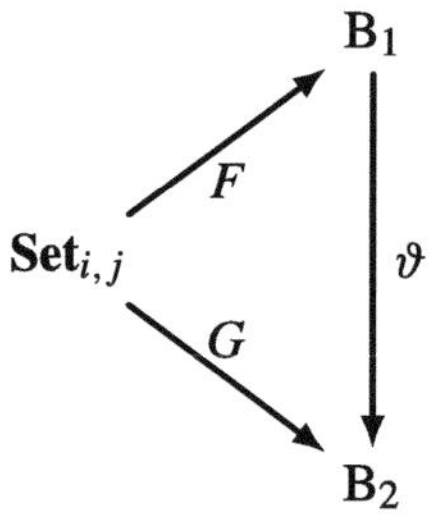

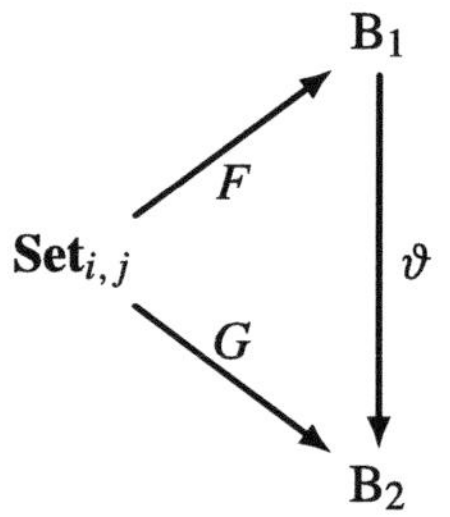

  where $\vartheta : F\,\mathbf{Set}_{ij}^{\mathrm{Coup}} \to G\,\mathbf{Set}_{ij}^{\mathrm{Coup}}$ for $i \neq j$ and $\vartheta : F\,\mathbf{Set}_{ii}^{\mathrm{Coup}} \to G\,\mathbf{Set}_{ii}^{\mathrm{Coup}}$ for $i = j$, and $\mathbf{Set}_{ij}^{\mathrm{Coup}}$ and $\mathbf{Set}_{ii}^{\mathrm{Coup}}$ are sets of coupling conditions, which are included in the objects $\mathbf{Set}_{i,j}$ and $\mathbf{Set}_{i,i}$, respectively.

Furthermore, model formulations, which are convertible on the left, convertible on the right, and convertible in coupling will be called *fully convertible*. Finally, coupled mathematical models which are instantiated by convertible model formulations will be called *convertible coupled models on the left*, *convertible coupled models on the right*, *convertible coupled models in coupling*, or *fully convertible coupled models*.

This definition underlines the fact, that the model formulation for a coupled model will not change, if models being coupled are converted to another formulations at first and then coupled, or if the coupled model is converted afterwards.

In the sequel, if no distinction between the type of convertibility for coupled mathematical models is necessary, we will use the general notion of convertible coupled mathematical models introduced in Definition 3.11, otherwise precise terms from Definition 3.12 will be used.

We immediately have the following corollary:

Corollary 3.5 *Convertible coupled mathematical models have the same complexity.*

It is worth to mention, that with the help of Definition 3.12, the notion of convertible coupled mathematical models can be connected to the different types of model complexity introduced in and around Definition 3.8, if such a specification is required.

The above discussion on convertibility of coupled mathematical models emphasises clearly, once again, the need for careful treatment of such models. In particular, it is important that categorical constructions for these models accurately reflect the way how models are created; otherwise the background of coupled mathematical models could be lost. Furthermore, it is also clearly visible from the constructions introduced above that the truly new aspects of coupled mathematical models are related to the coupling conditions. In fact, the most characterisations of coupled models are related to the individual mathematical models being coupled, as it could be expected from the modelling perspective.

3.3.4 Some Illustrative Examples for Coupled Models

Before introducing more refined categorical constructions for coupled mathematical models, let us provide some illustrative examples of using category theory-based modelling framework for description of coupled mathematical models. Moreover, based on our discussion in the previous subsections, we will provide examples for the cases when mathematical models from different categories and from the same category are coupled. In particular, we will describe category **Heat** containing several heat conduction models, and we will use this category to illustrate coupling of mathematical models from the same category. After that, models from the category **Heat** will be coupled with some models from the category **Elast**, which has been discussed already in the previous section.

3.3.4.1 Coupling of Mathematical Models from the Same Category

In this example, we consider the modelling of a heat conduction process in a domain Ω composed of two subdomains Ω_1 and Ω_2. This setting leads to the coupling of

Table 3.3 Sets of assumptions of heat conduction models

	Assumptions	$\mathbf{Set}_{\mathrm{UH}}$	$\mathbf{Set}_{\mathrm{NUH}}$	$\mathbf{Set}_{\mathrm{RH}}$	$\mathbf{Set}_{\mathrm{NURH}}$
1.	Heat propagates in a body according to Fourier's law	+	+	+	+
2.	The rate of change in internal heat energy per unit volume in the material is proportional to the rate of change of its temperature	+		+	
3.	A body is made of a homogeneous material	+		+	
4.	There is no radiative heat exchange with the surroundings according to the Stefan–Boltzmann law	+	+		
5.	External heat sources are allowed	+	+	+	+

heat conduction models used in different parts of the domain, and these models belong to the same category **Heat**.

Among various heat conduction models, we consider a heat conduction process in a uniform material, non-uniform material, in the presence of a radiative heat exchange with the surroundings, and the radiative heat exchange for the body made of a non-uniform material. The sets of assumptions corresponding to these heat conduction process are denoted as $\mathbf{Set}_{\mathrm{UH}}$, $\mathbf{Set}_{\mathrm{NUH}}$, $\mathbf{Set}_{\mathrm{RH}}$, and $\mathbf{Set}_{\mathrm{NURH}}$, respectively. These sets of assumptions are objects in the category **Heat**. At first, we describe the structure of this category, and after that, we will discuss how models from **Heat** can be coupled. Table 3.3 provides a list of sets of assumptions for the four models of heat conduction.

Model formulations of the sets of assumptions provided in Table 3.3 in the form of differential equations are given by:

$$\mathbf{Set}_{\mathrm{UH}} \overset{S}{\mapsto} c\rho\frac{\partial\theta}{\partial t} - \lambda\Delta\,\theta = F(\mathbf{x}, t) \qquad =: \mathbf{A},$$

$$\mathbf{Set}_{\mathrm{NUH}} \overset{S}{\mapsto} c\rho\frac{\partial\theta}{\partial t} - \nabla\cdot(\lambda\nabla\theta) = F(\mathbf{x}, t) \qquad =: \mathbf{B},$$

$$\mathbf{Set}_{\mathrm{RH}} \overset{S}{\mapsto} c\rho\frac{\partial\theta}{\partial t} - \lambda\Delta\,\theta + \mu\left(\theta^4 - \upsilon^4\right) = F(\mathbf{x}, t) \qquad =: \mathbf{C},$$

$$\mathbf{Set}_{\mathrm{NURH}} \overset{S}{\mapsto} c\rho\frac{\partial\theta}{\partial t} - \nabla\cdot(\lambda\nabla\theta) + \mu\left(\theta^4 - \upsilon^4\right) = F(\mathbf{x}, t) =: \mathbf{D},$$

where $\theta = \theta(\mathbf{x}, t)$ is the temperature, c is the heat capacity, λ is the thermal conductivity, ρ is the material density, μ is the Boltzmann constant, $\upsilon = \upsilon(\mathbf{x}, t)$ is the temperature of the surroundings, $F(\mathbf{x}, t)$ is the external heat source, $\mathbf{x} \in \mathbb{R}^3$ is the coordinate vector, $t \in \mathbb{R}_+$ is the time, and S are formalisation functors.

The structure of the category **Heat** is fully described by the following diagrams on the level of objects

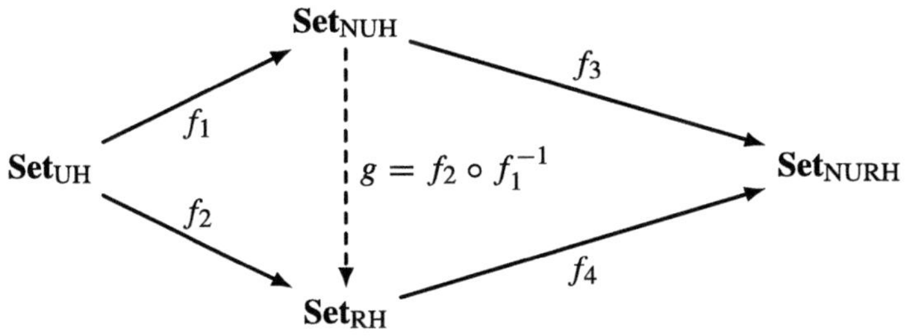

and models

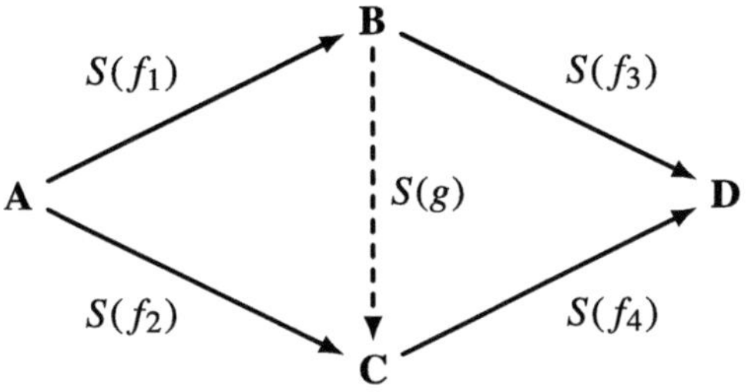

Here, the dashed arrow indicates that there is no direct morphism between the objects $\mathbf{Set}_{\mathrm{NUH}}$ and $\mathbf{Set}_{\mathrm{RH}}$, but it is still formally possible to compose the inverse morphism f_1^{-1}, implying simplification of a non-uniform material parameter λ to a constant value, and the morphism f_2 adding the influence of a radiative heat exchange with the surroundings.

We can summarise the following facts about the category **Heat**:

- **Heat** is a partially ordered category;
- The object $\mathbf{Set}_{\mathrm{UH}}$ is the initial object of this category;
- The object $\mathbf{Set}_{\mathrm{NURH}}$ is the terminal object of this category;
- **Heat** contains two totally ordered sub-categories:

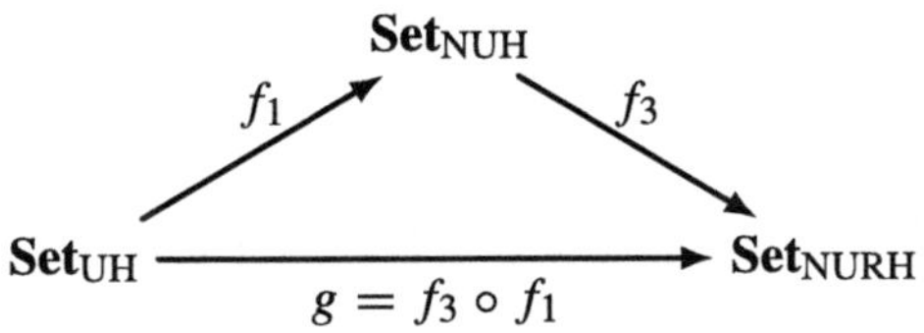

and

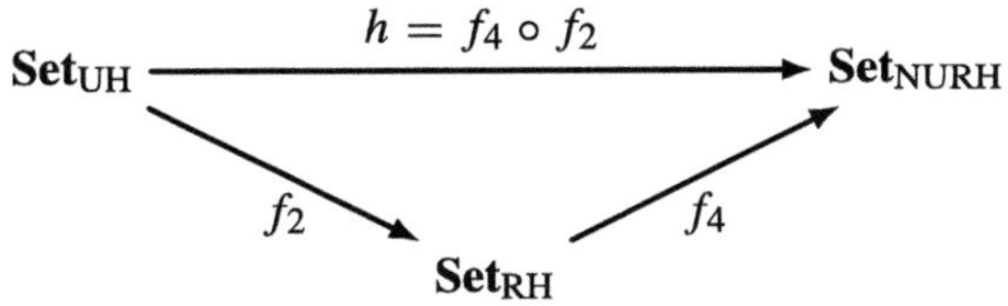

Evidently, we have the equivalence of morphism $f_3 \circ f_1 = f_4 \circ f_2$, or $g = h$, in other words, we have the pair of arrows:

$$\mathbf{Set}_{\mathrm{UH}} \underset{h = f_4 \circ f_2}{\overset{g = f_3 \circ f_1}{\rightrightarrows}} \mathbf{Set}_{\mathrm{NURH}}$$

This fact further implies that we can formally construct an *equaliser*, see Definition 2.14, consisting of an object and a universal arrow in **Heat**. It is important to recall, that the categories of mathematical models can be seen as special cases of a more general category **Sets**, where it holds that every subset $U \subseteq A$ is an equaliser for some pair of functions, see [1] for details. Hence, we can construct an equaliser in **Heat** by selecting any subset of $\mathbf{Set}_{\mathrm{UH}}$. Practically it means, that a universal arrow denoted by e in **Heat** takes any subset of basic assumptions $\mathbf{Set}_{\mathrm{UH}}$, which is not associated to any mathematical model of a heat conduction process, and extends it to the initial object of **Heat** and further satisfies the relation $g \circ e = h \circ e$.

After describing the structure of the category **Heat**, let us now discuss how models of this category can be coupled. A typical example for such a coupling, as we have mentioned already, is modelling of a heat conduction process in a domain composed of two subdomains, where different heat conduction models can be used, or the same model but with different material parameters:

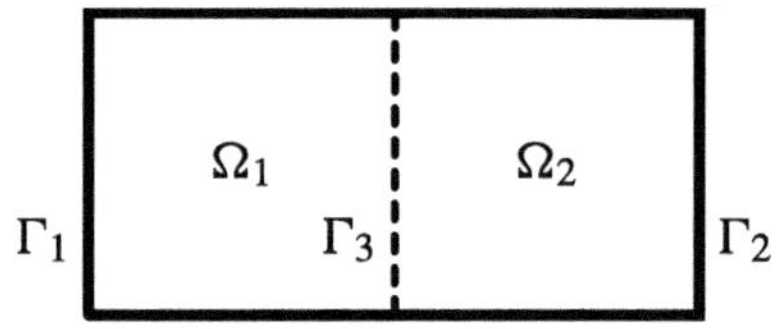

Here, Γ_3 is the coupling interface between two models of heat conduction used in subdomains Ω_1 and Ω_2.

We start constructing the coproduct category **HeatCoup** by using Definition 3.7, implying that we formally have

$$\mathbf{HeatCoup} = \mathbf{Heat} + \mathbf{Heat},$$

with objects of the following form:

$$\begin{aligned}
\mathbf{Set}_{\mathrm{UH,UH}} &:= \langle \mathbf{Set}_{\mathrm{UH}} + \mathbf{Set}_{\mathrm{UH}}, \mathbf{Set}^{\mathrm{Coup}}_{\mathrm{UH,UH}} \rangle, \\
\mathbf{Set}_{\mathrm{UH,NUH}} &:= \langle \mathbf{Set}_{\mathrm{UH}} + \mathbf{Set}_{\mathrm{NUH}}, \mathbf{Set}^{\mathrm{Coup}}_{\mathrm{UH,NUH}} \rangle, \\
\mathbf{Set}_{\mathrm{UH,RH}} &:= \langle \mathbf{Set}_{\mathrm{UH}} + \mathbf{Set}_{\mathrm{RH}}, \mathbf{Set}^{\mathrm{Coup}}_{\mathrm{UH,RH}} \rangle, \\
\mathbf{Set}_{\mathrm{UH,NURH}} &:= \langle \mathbf{Set}_{\mathrm{UH}} + \mathbf{Set}_{\mathrm{NURH}}, \mathbf{Set}^{\mathrm{Coup}}_{\mathrm{UH,NURH}} \rangle,
\end{aligned}$$

$$
\begin{aligned}
\mathbf{Set}_{\text{NUH,UH}} &:= \langle \mathbf{Set}_{\text{NUH}} + \mathbf{Set}_{\text{UH}}, \mathbf{Set}^{\text{Coup}}_{\text{NUH,UH}} \rangle, \\
\mathbf{Set}_{\text{NUH,NUH}} &:= \langle \mathbf{Set}_{\text{NUH}} + \mathbf{Set}_{\text{NUH}}, \mathbf{Set}^{\text{Coup}}_{\text{NUH,NUH}} \rangle, \\
\mathbf{Set}_{\text{NUH,RH}} &:= \langle \mathbf{Set}_{\text{NUH}} + \mathbf{Set}_{\text{RH}}, \mathbf{Set}^{\text{Coup}}_{\text{NUH,RH}} \rangle, \\
\mathbf{Set}_{\text{NUH,NURH}} &:= \langle \mathbf{Set}_{\text{NUH}} + \mathbf{Set}_{\text{NURH}}, \mathbf{Set}^{\text{Coup}}_{\text{NUH,NURH}} \rangle,
\end{aligned}
$$

$$
\begin{aligned}
\mathbf{Set}_{\text{RH,UH}} &:= \langle \mathbf{Set}_{\text{RH}} + \mathbf{Set}_{\text{UH}}, \mathbf{Set}^{\text{Coup}}_{\text{RH,UH}} \rangle, \\
\mathbf{Set}_{\text{RH,NUH}} &:= \langle \mathbf{Set}_{\text{RH}} + \mathbf{Set}_{\text{NUH}}, \mathbf{Set}^{\text{Coup}}_{\text{RH,NUH}} \rangle, \\
\mathbf{Set}_{\text{RH,RH}} &:= \langle \mathbf{Set}_{\text{RH}} + \mathbf{Set}_{\text{RH}}, \mathbf{Set}^{\text{Coup}}_{\text{RH,RH}} \rangle, \\
\mathbf{Set}_{\text{RH,NURH}} &:= \langle \mathbf{Set}_{\text{RH}} + \mathbf{Set}_{\text{NURH}}, \mathbf{Set}^{\text{Coup}}_{\text{RH,NURH}} \rangle,
\end{aligned}
$$

$$
\begin{aligned}
\mathbf{Set}_{\text{NURH,UH}} &:= \langle \mathbf{Set}_{\text{NURH}} + \mathbf{Set}_{\text{UH}}, \mathbf{Set}^{\text{Coup}}_{\text{NURH,UH}} \rangle, \\
\mathbf{Set}_{\text{NURH,NUH}} &:= \langle \mathbf{Set}_{\text{NURH}} + \mathbf{Set}_{\text{NUH}}, \mathbf{Set}^{\text{Coup}}_{\text{NURH,NUH}} \rangle, \\
\mathbf{Set}_{\text{NURH,RH}} &:= \langle \mathbf{Set}_{\text{NURH}} + \mathbf{Set}_{\text{RH}}, \mathbf{Set}^{\text{Coup}}_{\text{NURH,RH}} \rangle, \\
\mathbf{Set}_{\text{NURH,NURH}} &:= \langle \mathbf{Set}_{\text{NURH}} + \mathbf{Set}_{\text{NURH}}, \mathbf{Set}^{\text{Coup}}_{\text{NURH,NURH}} \rangle,
\end{aligned}
$$

where all possible combinations of models on the left and on the right have been taken into account. In total, we have $2^4 = 16$ objects in the category **HeatCoup**, which is described by the formula

$$
\natural\mathbf{ModCoup}_{i,i} = 2^{\natural\mathbf{Model}_i}, \tag{3.6}
$$

where $\natural$ denotes the number of objects in a category. It is necessary to make the remark:

Remark 3.14 Evidently, formula (3.6) is valid only in the case if all models associated with the objects in a category $\mathbf{Model}_i$ can be coupled among them, which is not necessarily always the case.

It is important to emphasise that objects such as for example $\mathbf{Set}_{\text{UH,NUH}}$ and $\mathbf{Set}_{\text{NUH,UH}}$ must to be considered separately. Although they use the same models of the heat conduction process, but these models are applied in different subdomains, which is essential information for constructing the complete coupled model. This fact is clearly indicated by the signature of these objects, and further underlines the advantage of working with coproducts in this context.

Next, we need to describe the coupling conditions used in the category **HeatCoup**. Because classical models of the heat conduction process are considered, heat propagates in the domain and its subdomains according to Fourier's law. This fact leads to two coupling conditions at the coupling interface Γ_3:

1. the temperature field must be continuous, and
2. the heat flux must be continuous,

which are valid for all considered heat conduction models, see [22]. Thus, the set of coupling conditions has the form

$$\mathbf{Set}_{i,i}^{\mathrm{Coup}} = \left\{ \begin{array}{l} \text{the temperature field must be continuous,} \\ \text{the heat flux must be continuous} \end{array} \right\},$$

where $i = \{\mathrm{UH}, \mathrm{NUH}, \mathrm{RH}, \mathrm{NURH}\}$. Further, the set of coupling conditions is formalised via a functorial mapping S as follows

$$\mathbf{Set}_{i,i}^{\mathrm{Coup}} \overset{S}{\mapsto} \left\{ \begin{array}{ll} [\theta] & = 0 \text{ on } \Gamma_3 \\ \left[\lambda \dfrac{\partial \theta}{\partial n} \right] & = 0 \text{ on } \Gamma_3 \end{array} \right.$$

To finish our discussion of this example, let us now list some additional facts about the category **HeatCoup**:

- All models associated with objects of **HeatCoup** are equal in coupling;
- The category **HeatCoup** is totally ordered in coupling (although it is trivial);
- The category **HeatCoup** is partially ordered on the left and on the right, as it has been discussed during construction of **Heat**.

3.3.4.2 Coupling of Mathematical Models from Different Categories

The example of coupling of mathematical models from different categories will be illustrated by discussing thermoelasticity models, which couple heat conduction models with models of linear elasticity. To this end, we introduce the category **ThermoElast**, containing sets of assumptions corresponding to various thermoelasticity models. However, before presenting the structure of this category, we first recall some basic concepts of thermoelasticity. For keeping presentation short, we will only provide the final forms of equations along with brief comments on their constructions. For a more detailed information on thermoelasticity, we refer the reader to [12, 14].

In the theory of linear thermoelasticity, additionally to mechanical strains, strains caused by the temperature change are considered. This implies, that the constitutive equations between stresses and strains are given by

$$\sigma_{ij} = 2\mu\varepsilon_{ij} + [\lambda\varepsilon_{kk} - (3\lambda + 2\mu)\alpha_T\Theta]\,\delta_{ij},$$

where $\Theta = T - T_0$, ($T_0 = \text{const.}$) with T being temperature, α_T is the thermal expansion coefficient, λ and μ are the Lamé coefficients of isothermal deformation for a fixed temperature $T = T_0$.

According to [12, 14], three general models of linear thermoelasticity can be distinguished:

1. coupled dynamic thermoelasticity model,
2. coupled quasi-static thermoelasticity model, and
3. decoupled dynamic thermoelasticity model, or, simply dynamic thermoelasticity model.

Let us list these three models given by partial differential equations:

- Coupled dynamic thermoelasticity model is given by the following system of differential equations:

$$\begin{cases} \mu\Delta\mathbf{u} + (\lambda+\mu)\operatorname{grad}\operatorname{div}\mathbf{u} - (3\lambda+2\mu)\alpha_T \operatorname{grad}\Theta + \mathbf{F} - \rho\ddot{\mathbf{u}} = 0, \\ \Delta\Theta - \dfrac{1}{a}\dot{\Theta} + \dfrac{w}{\lambda_q} - \dfrac{(3\lambda+2\mu)\alpha_T T_0}{\lambda_q}\operatorname{div}\mathbf{u} = 0, \end{cases}$$

  where λ_q is the heat conduction coefficient, w is the specific power of heat sources, ρ is the material density, $a = \dfrac{\lambda_q}{c_q}$ is the thermal conductivity with c_q being the heat capacity, and $\mathbf{F}$ is the vector of body forces.
- Coupled quasi-static thermoelasticity model is given by the following system of differential equations:

$$\begin{cases} \mu\Delta\mathbf{u} + (\lambda+\mu)\operatorname{grad}\operatorname{div}\mathbf{u} - (3\lambda+2\mu)\alpha_T \operatorname{grad}\Theta + \mathbf{F} = 0, \\ \Delta\Theta - \dfrac{1}{a}(1+\varepsilon)\Theta + \dfrac{w}{\lambda_q} = 0, \end{cases}$$

  where

$$\varepsilon = \frac{(3\lambda+2\mu)^2\alpha_T^2 T_0}{(\lambda+2\mu)c_q}$$

  is the coupling coefficient.
- Dynamic thermoelasticity model is given by the following system of differential equations:

$$\begin{cases} \mu\Delta\mathbf{u} + (\lambda+\mu)\operatorname{grad}\operatorname{div}\mathbf{u} - (3\lambda+2\mu)\alpha_T \operatorname{grad}\Theta + \mathbf{F} - \rho\ddot{\mathbf{u}} = 0, \\ \Delta\Theta - \dfrac{1}{a}\dot{\Theta} + \dfrac{w}{\lambda_q} = 0. \end{cases}$$

In fact, the difference between these three thermoelasticity models lies in the coupling. In essence, we have just described coupling complexity of thermoelasticity models. To make further constructions clearer, let us now formally consider three objects: $\mathbf{Set}_{\mathrm{CDT}}^{\mathrm{Coup}}$, $\mathbf{Set}_{\mathrm{CQST}}^{\mathrm{Coup}}$, and $\mathbf{Set}_{\mathrm{DT}}^{\mathrm{Coup}}$. These objects represent the coupling

Table 3.4 Sets of assumptions for coupling conditions between models of heat conduction and elasticity

	Assumptions	$\mathbf{Set}_{\mathrm{CDT}}^{\mathrm{Coup}}$	$\mathbf{Set}_{\mathrm{CQST}}^{\mathrm{Coup}}$	$\mathbf{Set}_{\mathrm{DT}}^{\mathrm{Coup}}$
1.	Linear form of constitutive equations is considered	+	+	+
2.	Heat conduction process affects mechanical deformations	+	+	+
3.	Mechanical deformations do not affect the temperature field		+	+
4.	Static process is considered		+	

conditions of the coupled dynamic thermoelasticity model, the coupled quasi-static thermoelasticity model, and the dynamic thermoelasticity model, respectively. Table 3.4 provides a list of assumptions for the coupling conditions for three thermoelasticity models.

It is also important to emphasise that, in this particular example, the coupling conditions are not affected by individual models of heat conduction and elasticity, as long as linear elasticity models are considered. This is clearly visible from Table 3.4. Consequently, when discussing the coupling of objects of **Heat** and **Elast**, we will omit the object $\mathbf{Set}_{\mathrm{NE}}$, which is related to the nonlinear model of elasticity. Additionally, as discussed above, for each coupled model associated with the objects of product category **ThermoElast** there are three sets of coupling conditions $\mathbf{Set}_{\mathrm{CDT}}^{\mathrm{Coup}}$, $\mathbf{Set}_{\mathrm{CQST}}^{\mathrm{Coup}}$, and $\mathbf{Set}_{\mathrm{DT}}^{\mathrm{Coup}}$, reflecting different levels of coupling complexity. Thus, the product category **ThermoElast** is formally given according to Definition 3.6 as follows

$$\mathbf{ThermoElast} = \mathbf{Elast} \times \mathbf{Heat}.$$

To demonstrate all objects of the category **ThermoElast**, we will examine three objects from the category **Elast**: namely $\mathbf{Set}_{\mathrm{LE}}$, $\mathbf{Set}_{\mathrm{ME}}$, and $\mathbf{Set}_{\mathrm{LEF}}$. For each of these objects, we will present all possible objects of category **ThermoElast**. Hence, we have the following combinations:

- Object $\mathbf{Set}_{\mathrm{LE}}$:

$$\begin{aligned}
\mathbf{Set}_{\mathrm{LE,UH}}^{\mathrm{CDT}} &:= \langle \mathbf{Set}_{\mathrm{LE}} \cup \mathbf{Set}_{\mathrm{UH}}, \mathbf{Set}_{\mathrm{CDT}}^{\mathrm{Coup}} \rangle, \\
\mathbf{Set}_{\mathrm{LE,UH}}^{\mathrm{CQST}} &:= \langle \mathbf{Set}_{\mathrm{LE}} \cup \mathbf{Set}_{\mathrm{UH}}, \mathbf{Set}_{\mathrm{CQST}}^{\mathrm{Coup}} \rangle, \\
\mathbf{Set}_{\mathrm{LE,UH}}^{\mathrm{DT}} &:= \langle \mathbf{Set}_{\mathrm{LE}} \cup \mathbf{Set}_{\mathrm{UH}}, \mathbf{Set}_{\mathrm{DT}}^{\mathrm{Coup}} \rangle,
\end{aligned}$$

$$\begin{aligned}
\mathbf{Set}_{\mathrm{LE,NUH}}^{\mathrm{CDT}} &:= \langle \mathbf{Set}_{\mathrm{LE}} \cup \mathbf{Set}_{\mathrm{NUH}}, \mathbf{Set}_{\mathrm{CDT}}^{\mathrm{Coup}} \rangle, \\
\mathbf{Set}_{\mathrm{LE,NUH}}^{\mathrm{CQST}} &:= \langle \mathbf{Set}_{\mathrm{LE}} \cup \mathbf{Set}_{\mathrm{NUH}}, \mathbf{Set}_{\mathrm{CQST}}^{\mathrm{Coup}} \rangle, \\
\mathbf{Set}_{\mathrm{LE,NUH}}^{\mathrm{DT}} &:= \langle \mathbf{Set}_{\mathrm{LE}} \cup \mathbf{Set}_{\mathrm{NUH}}, \mathbf{Set}_{\mathrm{DT}}^{\mathrm{Coup}} \rangle,
\end{aligned}$$

$$\begin{aligned}
\mathbf{Set}^{\mathrm{CDT}}_{\mathrm{LE,RH}} &:= \langle \mathbf{Set}_{\mathrm{LE}} \cup \mathbf{Set}_{\mathrm{RH}}, \mathbf{Set}^{\mathrm{Coup}}_{\mathrm{CDT}} \rangle, \\
\mathbf{Set}^{\mathrm{CQST}}_{\mathrm{LE,RH}} &:= \langle \mathbf{Set}_{\mathrm{LE}} \cup \mathbf{Set}_{\mathrm{RH}}, \mathbf{Set}^{\mathrm{Coup}}_{\mathrm{CQST}} \rangle, \\
\mathbf{Set}^{\mathrm{DT}}_{\mathrm{LE,RH}} &:= \langle \mathbf{Set}_{\mathrm{LE}} \cup \mathbf{Set}_{\mathrm{RH}}, \mathbf{Set}^{\mathrm{Coup}}_{\mathrm{DT}} \rangle,
\end{aligned}$$

$$\begin{aligned}
\mathbf{Set}^{\mathrm{CDT}}_{\mathrm{LE,NURH}} &:= \langle \mathbf{Set}_{\mathrm{LE}} \cup \mathbf{Set}_{\mathrm{NURH}}, \mathbf{Set}^{\mathrm{Coup}}_{\mathrm{CDT}} \rangle, \\
\mathbf{Set}^{\mathrm{CQST}}_{\mathrm{LE,NURH}} &:= \langle \mathbf{Set}_{\mathrm{LE}} \cup \mathbf{Set}_{\mathrm{NURH}}, \mathbf{Set}^{\mathrm{Coup}}_{\mathrm{CQST}} \rangle, \\
\mathbf{Set}^{\mathrm{DT}}_{\mathrm{LE,NURH}} &:= \langle \mathbf{Set}_{\mathrm{LE}} \cup \mathbf{Set}_{\mathrm{NURH}}, \mathbf{Set}^{\mathrm{Coup}}_{\mathrm{DT}} \rangle,
\end{aligned}$$

- Object $\mathbf{Set}_{\mathrm{ME}}$:

$$\begin{aligned}
\mathbf{Set}^{\mathrm{CDT}}_{\mathrm{ME,UH}} &:= \langle \mathbf{Set}_{\mathrm{ME}} \cup \mathbf{Set}_{\mathrm{UH}}, \mathbf{Set}^{\mathrm{Coup}}_{\mathrm{CDT}} \rangle, \\
\mathbf{Set}^{\mathrm{CQST}}_{\mathrm{ME,UH}} &:= \langle \mathbf{Set}_{\mathrm{ME}} \cup \mathbf{Set}_{\mathrm{UH}}, \mathbf{Set}^{\mathrm{Coup}}_{\mathrm{CQST}} \rangle, \\
\mathbf{Set}^{\mathrm{DT}}_{\mathrm{ME,UH}} &:= \langle \mathbf{Set}_{\mathrm{ME}} \cup \mathbf{Set}_{\mathrm{UH}}, \mathbf{Set}^{\mathrm{Coup}}_{\mathrm{DT}} \rangle,
\end{aligned}$$

$$\begin{aligned}
\mathbf{Set}^{\mathrm{CDT}}_{\mathrm{ME,NUH}} &:= \langle \mathbf{Set}_{\mathrm{ME}} \cup \mathbf{Set}_{\mathrm{NUH}}, \mathbf{Set}^{\mathrm{Coup}}_{\mathrm{CDT}} \rangle, \\
\mathbf{Set}^{\mathrm{CQST}}_{\mathrm{ME,NUH}} &:= \langle \mathbf{Set}_{\mathrm{ME}} \cup \mathbf{Set}_{\mathrm{NUH}}, \mathbf{Set}^{\mathrm{Coup}}_{\mathrm{CQST}} \rangle, \\
\mathbf{Set}^{\mathrm{DT}}_{\mathrm{ME,NUH}} &:= \langle \mathbf{Set}_{\mathrm{ME}} \cup \mathbf{Set}_{\mathrm{NUH}}, \mathbf{Set}^{\mathrm{Coup}}_{\mathrm{DT}} \rangle,
\end{aligned}$$

$$\begin{aligned}
\mathbf{Set}^{\mathrm{CDT}}_{\mathrm{ME,RH}} &:= \langle \mathbf{Set}_{\mathrm{ME}} \cup \mathbf{Set}_{\mathrm{RH}}, \mathbf{Set}^{\mathrm{Coup}}_{\mathrm{CDT}} \rangle, \\
\mathbf{Set}^{\mathrm{CQST}}_{\mathrm{ME,RH}} &:= \langle \mathbf{Set}_{\mathrm{ME}} \cup \mathbf{Set}_{\mathrm{RH}}, \mathbf{Set}^{\mathrm{Coup}}_{\mathrm{CQST}} \rangle, \\
\mathbf{Set}^{\mathrm{DT}}_{\mathrm{ME,RH}} &:= \langle \mathbf{Set}_{\mathrm{ME}} \cup \mathbf{Set}_{\mathrm{RH}}, \mathbf{Set}^{\mathrm{Coup}}_{\mathrm{DT}} \rangle,
\end{aligned}$$

$$\begin{aligned}
\mathbf{Set}^{\mathrm{CDT}}_{\mathrm{ME,NURH}} &:= \langle \mathbf{Set}_{\mathrm{ME}} \cup \mathbf{Set}_{\mathrm{NURH}}, \mathbf{Set}^{\mathrm{Coup}}_{\mathrm{CDT}} \rangle, \\
\mathbf{Set}^{\mathrm{CQST}}_{\mathrm{ME,NURH}} &:= \langle \mathbf{Set}_{\mathrm{ME}} \cup \mathbf{Set}_{\mathrm{NURH}}, \mathbf{Set}^{\mathrm{Coup}}_{\mathrm{CQST}} \rangle, \\
\mathbf{Set}^{\mathrm{DT}}_{\mathrm{ME,NURH}} &:= \langle \mathbf{Set}_{\mathrm{ME}} \cup \mathbf{Set}_{\mathrm{NURH}}, \mathbf{Set}^{\mathrm{Coup}}_{\mathrm{DT}} \rangle,
\end{aligned}$$

- Object $\mathbf{Set}_{\mathrm{LEF}}$:

$$\begin{aligned}
\mathbf{Set}^{\mathrm{CDT}}_{\mathrm{LEF,UH}} &:= \langle \mathbf{Set}_{\mathrm{LEF}} \cup \mathbf{Set}_{\mathrm{UH}}, \mathbf{Set}^{\mathrm{Coup}}_{\mathrm{CDT}} \rangle, \\
\mathbf{Set}^{\mathrm{CQST}}_{\mathrm{LEF,UH}} &:= \langle \mathbf{Set}_{\mathrm{LEF}} \cup \mathbf{Set}_{\mathrm{UH}}, \mathbf{Set}^{\mathrm{Coup}}_{\mathrm{CQST}} \rangle, \\
\mathbf{Set}^{\mathrm{DT}}_{\mathrm{LEF,UH}} &:= \langle \mathbf{Set}_{\mathrm{LEF}} \cup \mathbf{Set}_{\mathrm{UH}}, \mathbf{Set}^{\mathrm{Coup}}_{\mathrm{DT}} \rangle,
\end{aligned}$$

$$\begin{aligned}
\mathbf{Set}^{\mathrm{CDT}}_{\mathrm{LEF,NUH}} &:= \langle \mathbf{Set}_{\mathrm{LEF}} \cup \mathbf{Set}_{\mathrm{NUH}}, \mathbf{Set}^{\mathrm{Coup}}_{\mathrm{CDT}} \rangle, \\
\mathbf{Set}^{\mathrm{CQST}}_{\mathrm{LEF,NUH}} &:= \langle \mathbf{Set}_{\mathrm{LEF}} \cup \mathbf{Set}_{\mathrm{NUH}}, \mathbf{Set}^{\mathrm{Coup}}_{\mathrm{CQST}} \rangle, \\
\mathbf{Set}^{\mathrm{DT}}_{\mathrm{LEF,NUH}} &:= \langle \mathbf{Set}_{\mathrm{LEF}} \cup \mathbf{Set}_{\mathrm{NUH}}, \mathbf{Set}^{\mathrm{Coup}}_{\mathrm{DT}} \rangle,
\end{aligned}$$

$$\begin{aligned}
\mathbf{Set}^{\mathrm{CDT}}_{\mathrm{LEF,RH}} &:= \langle \mathbf{Set}_{\mathrm{LEF}} \cup \mathbf{Set}_{\mathrm{RH}}, \mathbf{Set}^{\mathrm{Coup}}_{\mathrm{CDT}} \rangle, \\
\mathbf{Set}^{\mathrm{CQST}}_{\mathrm{LEF,RH}} &:= \langle \mathbf{Set}_{\mathrm{LEF}} \cup \mathbf{Set}_{\mathrm{RH}}, \mathbf{Set}^{\mathrm{Coup}}_{\mathrm{CQST}} \rangle, \\
\mathbf{Set}^{\mathrm{DT}}_{\mathrm{LEF,RH}} &:= \langle \mathbf{Set}_{\mathrm{LEF}} \cup \mathbf{Set}_{\mathrm{RH}}, \mathbf{Set}^{\mathrm{Coup}}_{\mathrm{DT}} \rangle,
\end{aligned}$$

$$\begin{aligned}
\mathbf{Set}^{\mathrm{CDT}}_{\mathrm{LEF,NURH}} &:= \langle \mathbf{Set}_{\mathrm{LEF}} \cup \mathbf{Set}_{\mathrm{NURH}}, \mathbf{Set}^{\mathrm{Coup}}_{\mathrm{CDT}} \rangle, \\
\mathbf{Set}^{\mathrm{CQST}}_{\mathrm{LEF,NURH}} &:= \langle \mathbf{Set}_{\mathrm{LEF}} \cup \mathbf{Set}_{\mathrm{NURH}}, \mathbf{Set}^{\mathrm{Coup}}_{\mathrm{CQST}} \rangle, \\
\mathbf{Set}^{\mathrm{DT}}_{\mathrm{LEF,NURH}} &:= \langle \mathbf{Set}_{\mathrm{LEF}} \cup \mathbf{Set}_{\mathrm{NURH}}, \mathbf{Set}^{\mathrm{Coup}}_{\mathrm{DT}} \rangle,
\end{aligned}$$

In total, we have $3 \cdot 4 \cdot 4 = 36$ objects in the category **ThermoElast**. This fact is described by the formula

$$\natural\mathbf{ModCoup}_{i,j} = \natural\mathbf{Model}_i \cdot \natural\mathbf{Model}_j \cdot \natural\mathbf{Set}^{\mathrm{Coup}}_{i,j}, \tag{3.7}$$

where we extend the use of $\natural$ to give not only the number of objects in a category, but the number of different coupling conditions as well. Similar to formula (3.6), formula (3.7) is valid only in the case if all models associated with the objects in a categories $\mathbf{Model}_i$ and $\mathbf{Model}_j$ can be coupled among them, which is not necessarily always the case.

Remark 3.15 It is important to note, that it is not possible to provide formulae analogous to (3.6)–(3.7) for the general case, because of too many possibilities which would need to be addressed then by such formulae. However, formulae (3.6)–(3.7) can be interpreted as an upper bound for the number of coupled models in the case if all models from a category $\mathbf{Model}_i$ can be coupled with all models from category $\mathbf{Model}_j$ and all coupling conditions are applicable to every possible model pair.

A diagrammatic representation of the structure of **ThermoElast** is tricky because of many combinations of models and coupling conditions. Therefore, let us at first provide a diagram for the coupling conditions presented in Table 3.4:

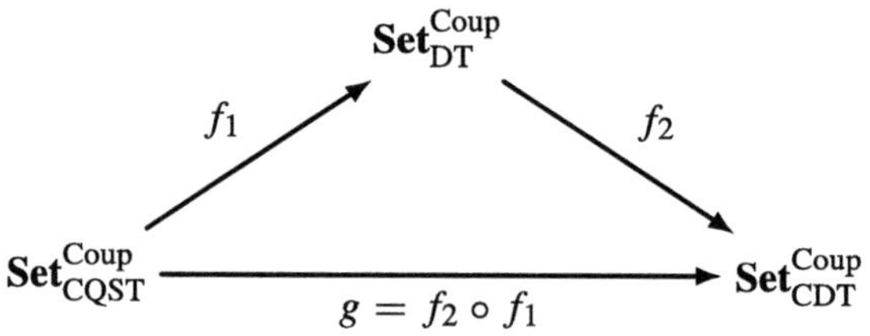

Thus, the set of coupling conditions $\mathbf{Set}^{\mathrm{Coup}}_{\mathrm{CDT}}$ provides models with the highest coupling complexity. Therefore, in any triple of models $\mathbf{Set}^{\mathrm{CDT}}_{i,j}$, $\mathbf{Set}^{\mathrm{CQST}}_{i,j}$, and $\mathbf{Set}^{\mathrm{DT}}_{i,j}$ with $i = \{\mathrm{LE, ME, LEF}\}$, $j = \{\mathrm{UH, NUH, RH, NURH}\}$, the above diagram holds for the coupling complexity.

Next, let us examine two combinations of models:

1. $\mathbf{Set}^{\text{CDT}}_{\text{LE,UH}}$, $\mathbf{Set}^{\text{CDT}}_{\text{LE,NUH}}$, $\mathbf{Set}^{\text{CDT}}_{\text{LE,RH}}$, and $\mathbf{Set}^{\text{CDT}}_{\text{LE,NURH}}$;
2. $\mathbf{Set}^{\text{CDT}}_{\text{LE,UH}}$, $\mathbf{Set}^{\text{CDT}}_{\text{ME,UH}}$, and $\mathbf{Set}^{\text{CDT}}_{\text{LEF,UH}}$.

For these two combinations of models, we have the following diagrams:

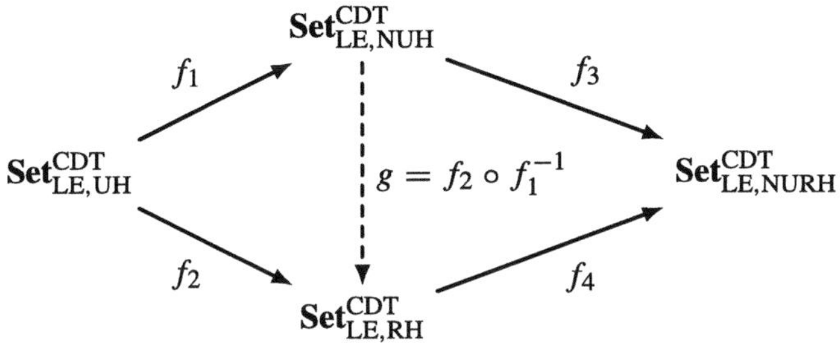

and

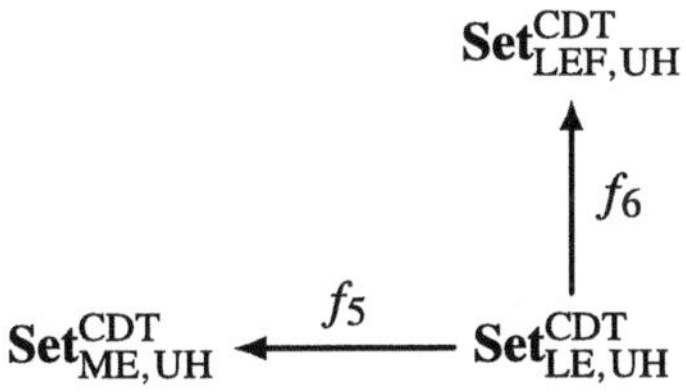

correspondingly. These diagrams indicate that both combinations of models differ in their base complexity. More precisely, the first combination of models differs in the right-base complexity but being equal in the left-base complexity. Conversely, the second combination of models differs in the left-base complexity but being equal in the right-base complexity.

Evidently, by using projections π_1 and π_2, we would end up in the category **Elast** (in fact, its subcategory, since the object $\mathbf{Set}_{\text{NE}}$ has been omitted) and in category **Heat**, respectively. In this way, we decouple the coupled models using these projection functors. All the results presented for these two categories do still hold in the case of their coupling. Similar diagrams can be constructed for any other combinations of objects from **ThermoElast**.

Let us now summarise some facts about the structure of the category **ThermoElast**:

- **ThermoElast** is a totally ordered category in coupling;
- **ThermoElast** is a partially ordered category on the left;
- **ThermoElast** is a partially ordered category on the right;
- The objects of the form $\mathbf{Set}^{\text{CDT}}_{i,j}$ are the most complex objects w.r.t. coupling. There are 12 of these objects and they are different in the base complexity;
- The objects of the form $\mathbf{Set}^{\text{CQST}}_{i,j}$ are the simplest objects w.r.t. coupling. There are 12 of these objects and they are different in the base complexity.

3.3.5 *Extension to the Coupling of Models From More than Two Categories*

Given that modern engineering and physics problems are often multiphysics in nature and, therefore, often require the coupling of models from more than two different fields, it is necessary to extend the results presented in the previous subsections to address the coupling of models from multiple categories. This coupling process is essentially an *upscale* of the coupling of two categories, and it is based on the concept of the lattice of theories proposed in [21], which we previously discussed in Sect. 2.3. Accordingly, we will focus on describing the general idea of this construction and highlight the main differences compared to the case of coupling mathematical models from only two categories.

Arguably, the most straightforward way of extending Definitions 3.6 and 3.7 to the case of coupling mathematical models from several categories is to couple them in a step-by-step manner. Consequently, the categories being coupled may themselves be categories of coupled mathematical models, and, thus, Definitions 3.6–3.7 must be upscaled to this setting. While this upscaling is straightforward, the increased number of possibilities for coupling models—either from different or from the same category—introduces technical details that worth further discussion.

From now on, we will address coupling of three categories, because an extension to more categories is completely analogous. We will denote categories of mathematical models to be coupled as $\mathbf{Model}_i$, $\mathbf{Model}_j$, and $\mathbf{Model}_k$. Additionally, we will make a convention that at first categories $\mathbf{Model}_i$ and $\mathbf{Model}_j$ are coupled, and then category $\mathbf{Model}_k$ is "added" to the result of this coupling.

For an easier extension to the case of more categories, we will also use the convention that order of indices in $\mathbf{ModCoup}_{i,j,k}$ corresponds to the order in which categories were coupled. For example, $\mathbf{ModCoup}_{i,j,k}$ means exactly that categories $\mathbf{Model}_i$ and $\mathbf{Model}_j$ are coupled first, and then the result of this coupling is coupled with $\mathbf{Model}_k$; while $\mathbf{ModCoup}_{i,k,j}$ means that categories $\mathbf{Model}_i$ and $\mathbf{Model}_k$ are coupled first.

To this end, we need to consider several cases for coupling three categories of mathematical models:

1. $i \neq j \neq k$—coupling of mathematical models from three different categories;
2. $i = j \neq k$—coupling of mathematical models, when two models belong to one category and the third model belongs to a different category;
3. $i = j = k$—coupling of three mathematical models from the same category.

Furthermore, for coupling of $\mathbf{Model}_i$ and $\mathbf{Model}_j$ we will use directly Definitions 3.6–3.7, because the main point of our discussion here is relations of i and j to k and its influence on the final structure of the category of coupled mathematical models.

For providing a better readability, we will present results for the three cases in the form of separate definitions. For the first case, we have:

Definition 3.13 (Case $i \neq j \neq k$) For $i \neq j \neq k$ coupling of the category $\mathbf{ModCoup}_{i,j}$ with the $\mathbf{Model}_k$ is a product category, formally written as

$$\mathbf{ModCoup}_{i,j,k} = \mathbf{ModCoup}_{i,j} \times \mathbf{Model}_k,$$

with projections functors:

$$\mathbf{ModCoup}_{i,j} \xleftarrow{\pi_1} \mathbf{ModCoup}_{i,j} \times \mathbf{Model}_k \xrightarrow{\pi_2} \mathbf{Model}_k.$$

Further, for all objects of $\mathbf{ModCoup}_{i,j,k}$, denoted by $\mathbf{Set}_{i,j,k}$, the following assumptions hold:

1. sets of assumptions $\mathbf{Set}_{i,j,k}$ are constructed formally as follows:

$$\mathbf{Set}_{i,j,k} := \langle \mathbf{Set}_i \cup \mathbf{Set}_j \cup \mathbf{Set}_k, \mathbf{Set}_{ij}^{\text{Coup}} \cup \mathbf{Set}_{ijk}^{\text{Coup}} \rangle,$$

   where the set $\mathbf{Set}_{ijk}^{\text{Coup}}$ represents the coupling conditions, which are necessary to couple models from $\mathbf{ModCoup}_{i,j}$ with models from $\mathbf{Model}_k$. Moreover, the following relations must be satisfied:

$$(a)\ \left(\mathbf{Set}_i \cup \mathbf{Set}_j \cup \mathbf{Set}_k\right) \subset \mathbf{Set}_{i,j,k}^{\text{U}}; \quad (b)\ \mathbf{Set}_{i,j,k}^{\text{U}} \not\subset \left(\mathbf{Set}_i \cup \mathbf{Set}_j \cup \mathbf{Set}_k\right),$$

   where $\mathbf{Set}_{i,j,k}^{\text{U}} = \mathbf{Set}_i \cup \mathbf{Set}_j \cup \mathbf{Set}_k \cup \mathbf{Set}_{ij}^{\text{Coup}} \cup \mathbf{Set}_{ijk}^{\text{Coup}}$;
2. each $\mathbf{Set}_{i,j,k}$ has the corresponding formalisation mapping S to the respective coupled mathematical model.

As we see, this definition is a straightforward extension of Definition 3.6. It is important to underline, that formally we have

$$\mathbf{ModCoup}_{i,j,k} \neq \mathbf{ModCoup}_{j,k,i} \neq \mathbf{ModCoup}_{i,k,j}$$

due to construction. Practically it means, that it is necessary to keep track of the order in which models are coupled, because different structures of categories would arise.

Let us now consider the second case:

Definition 3.14 (Case $i = j \neq k$) For $i = j \neq k$ coupling of the category $\mathbf{ModCoup}_{i,i}$ with the $\mathbf{Model}_k$ is a product category, formally written as

$$\mathbf{ModCoup}_{i,i,k} = \mathbf{ModCoup}_{i,i} \times \mathbf{Model}_k,$$

with projections functors:

$$\mathbf{ModCoup}_{i,i} \xleftarrow{\pi_1} \mathbf{ModCoup}_{i,i} \times \mathbf{Model}_k \xrightarrow{\pi_2} \mathbf{Model}_k.$$

Further, for all objects of $\mathbf{ModCoup}_{i,i,k}$, denoted by $\mathbf{Set}_{i,i,k}$, the following assumptions hold:

1. sets of assumptions $\mathbf{Set}_{i,i,k}$ are constructed formally as follows:

$$\mathbf{Set}_{i,i,k} := \langle \mathbf{Set}_i + \mathbf{Set}_i, \mathbf{Set}_k, \mathbf{Set}_{ii}^{\mathrm{Coup}} \cup \mathbf{Set}_{iik}^{\mathrm{Coup}} \rangle,$$

   where the set $\mathbf{Set}_{iik}^{\mathrm{Coup}}$ represents the coupling conditions, which are necessary to couple models from $\mathbf{ModCoup}_{i,i}$ with models from $\mathbf{Model}_k$. Moreover, the following relations must be satisfied:

$$(a)\ (\mathbf{Set}_i \cup \mathbf{Set}_k) \subset \mathbf{Set}_{i,i,k}^{\mathrm{U}}; \quad (b)\ \mathbf{Set}_{i,i,k}^{\mathrm{U}} \not\subset (\mathbf{Set}_i \cup \mathbf{Set}_k),$$

   where $\mathbf{Set}_{i,i,k}^{\mathrm{U}} = \mathbf{Set}_i \cup \mathbf{Set}_k \cup \mathbf{Set}_{ii}^{\mathrm{Coup}} \cup \mathbf{Set}_{iik}^{\mathrm{Coup}}$;
2. each $\mathbf{Set}_{i,i,k}$ has the corresponding formalisation mapping S to the respective coupled mathematical model.

The case $i = j \neq k$ creates an interesting "mixture" of product and coproduct categories, which of course comes naturally from the construction. It is also necessary to underline, that if we considered another order of coupling, namely i, k, i then the category $\mathbf{ModCoup}_{i,k}$ would be a product category, and the objects of $\mathbf{ModCoup}_{i,k,i}$ would be of the form

$$\mathbf{Set}_{i,k,i} := \langle (\mathbf{Set}_i \cup \mathbf{Set}_k) + \mathbf{Set}_i, \mathbf{Set}_{ik}^{\mathrm{Coup}} \cup \mathbf{Set}_{iki}^{\mathrm{Coup}} \rangle,$$

where the set of coupling conditions $\mathbf{Set}_{iki}^{\mathrm{Coup}}$ is in fact the same as $\mathbf{Set}_{iik}^{\mathrm{Coup}}$.

Finally, we address the third case:

Definition 3.15 (Case $i = j = k$) For $i = j = k$ coupling of the category $\mathbf{ModCoup}_{i,i}$ with the $\mathbf{Model}_i$ is a coproduct category, formally written as

$$\mathbf{ModCoup}_{i,i,i} = \mathbf{ModCoup}_{i,i} + \mathbf{Model}_i,$$

with injections:

$$\mathbf{Model}_{i,i} \xrightarrow{i_1} \mathbf{Model}_{i,i} + \mathbf{Model}_i \xleftarrow{i_2} \mathbf{Model}_i.$$

Moreover, for all objects of $\mathbf{ModCoup}_{i,i,i}$, denoted by $\mathbf{Set}_{i,i,i}$, the following assumptions hold:

1. sets of assumptions of a coupled mathematical models $\mathbf{Set}_{i,i,i}$ are constructed formally as follows:

$$\mathbf{Set}_{i,i,i} := \langle \mathbf{Set}_i + \mathbf{Set}_i + \mathbf{Set}_i, \mathbf{Set}_{ii}^{\mathrm{Coup}} \rangle,$$

where the set $\mathbf{Set}_{ii}^{\mathrm{Coup}} \neq \emptyset$ represents the coupling conditions, which are necessary to couple models from $\mathbf{Model}_i$;

2. each $\mathbf{Set}_{i,i,i}$ has the corresponding formalisation mapping S to the respective coupled mathematical model.

Similar to the first case, this definition is a straightforward extension of Definition 3.7. Additionally, in contrast to the first case, we do not have any extra coupling conditions to couple the third model, because the set $\mathbf{Set}_{ii}^{\mathrm{Coup}}$ contains already all coupling conditions, which are necessary to couple models from $\mathbf{Model}_i$. This, of course also implies, that the order in which models are coupled is not relevant for the structure of the category $\mathbf{ModCoup}_{i,i,i}$.

Definitions 3.13–3.15 present the general idea of how models from more than two categories can be coupled. Clearly, this construction can be naturally extended to the coupling of models from n categories. To provide an idea how such a coupling can be realised in practice, we will adapt the concept of the lattice of theories, see again [21].

As mentioned earlier, the coupling of multiple categories of mathematical models should be done in a step-by-step manner. However, even in that case there are several possibilities to approach the coupling. For simplicity, let us now consider five categories of mathematical models $\mathbf{Model}_i$, $i = 1, \ldots, 5$, and demonstrate how they can be coupled to construct a global model. Figure 3.2 presents a lattice for the coupling of models from these five categories. Each arrow in this lattice is a functor mapping between categories, which is interpreted as an expansion operation, and hence marked by $\mathbf{E}$. In this context, the *expansion of a mathematical model* is a process of coupling the model with another, and, thus, expanding the range of phenomena describable by this model.

$$
\begin{array}{ccccc}
\mathbf{Model}_5 & \xrightarrow{\mathbf{E}} & \mathbf{ModCoup}_{1,2,3,4,5} & & \\
 & & \uparrow \mathbf{E} & & \\
 & & \mathbf{ModCoup}_{1,2,3,4} & \xleftarrow{\mathbf{E}} & \mathbf{Model}_4 \\
 & & \uparrow \mathbf{E} & & \\
\mathbf{Model}_3 & \xrightarrow{\mathbf{E}} & \mathbf{ModCoup}_{1,2,3} & & \\
 & & \uparrow \mathbf{E} & & \\
 & & \mathbf{ModCoup}_{1,2} & \xleftarrow{\mathbf{E}} & \mathbf{Model}_2 \\
 & & \uparrow \mathbf{E} & & \\
 & & \mathbf{Model}_1 & &
\end{array}
$$

Fig. 3.2 A lattice of categories of mathematical models. Each arrows is a functor mapping, that expands a mathematical model to describe more physical phenomena

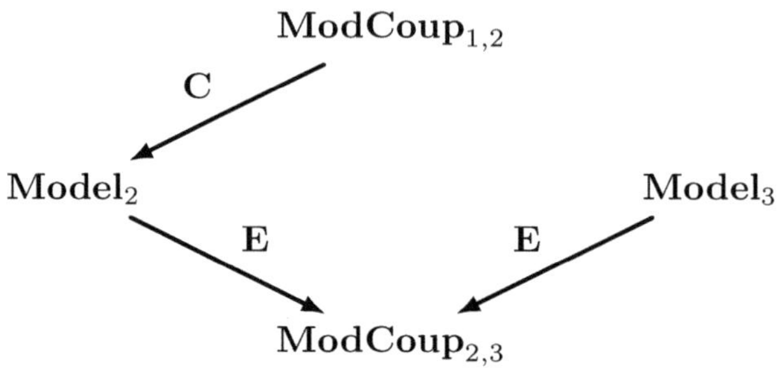

Fig. 3.3 Revision on a lattice of categories of mathematical models

By reversing the arrows in Fig. 3.2, the *contraction of a mathematical model* is obtained, meaning that we restrict the range of phenomena describable by a model. Combining this fact with the previous discussions, it can be stated, that the coupling of models is an expansion operation, while *decoupling* is a contraction operation.

Figure 3.3 illustrates how the revision of mathematical models can be interpreted within the category theory-based modelling framework. The idea here is to start with a category of coupled mathematical models and "extract" one category of mathematical models to couple it with another category. Thus, similar to the lattice of theories concept, the *revision of a mathematical model* is a composition of expansion and contraction $\mathbf{R} := \mathbf{E} \circ \mathbf{C}$. In other words, the revision of a mathematical model is an operation which discards some of the physical phenomena describable by the model (contraction) and adds new phenomena (expansion).

Let us make the following remark:

Remark 3.16 It is necessary to make clear that expansion $\mathbf{E}$ and contraction $\mathbf{C}$ operations can be directly described in terms of the categorical constructions for coupled models discussed in this section. In this context, the adaptation of the lattice of theories concept can be seen as a schematic representation of the coupling of models from multiple categories, without requiring immediate formal definitions of objects in all possible categories of coupled mathematical models.

Addressing the analogy in the context of category theory-based modelling framework is a bit more tricky, because, unlike in the lattice of theories, we work with categories of mathematical models, which are essentially the same "objects". Nonetheless, it is still possible to introduce analogy here, if we make our constructions more restrictive. Namely, let us consider a family of formalisation functors S_i, $i = 1, 2, \ldots$, which are connected via natural transformations ϑ_i, $i = 1, 2, \ldots$. If we now fix exactly one formalisation functor S_1 for a lattice of categories of mathematical models, then the *analogy of mathematical models* would be the action of a natural transformation ϑ_1 on this functor which translates one formalism, chosen to instantiate sets of assumptions, to another formalism. This perspective might be indeed convenient in some practical applications, because here we have a clear distinction between for example mathematical models in

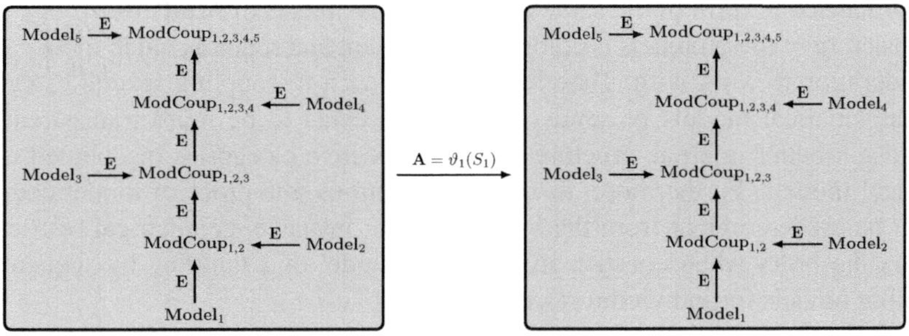

Fig. 3.4 The analogy of a lattice of categories of mathematical models, which is realised by a natural transformation ϑ_1 acting on formalisation functor S_1

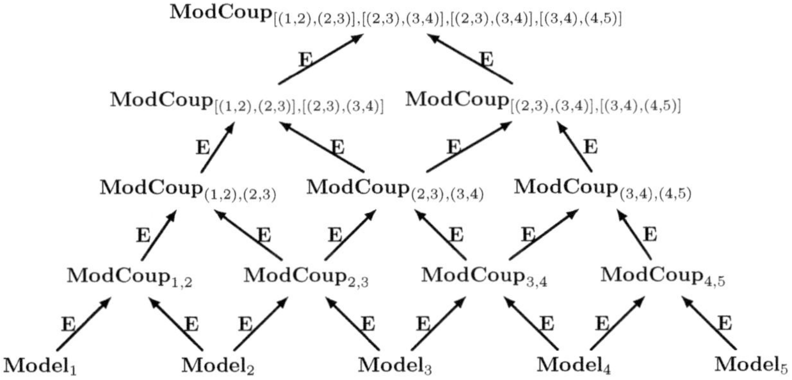

Fig. 3.5 An alternative approach to constructing lattices of categories of mathematical models. Indices indicate, that the internal structure of these categories is much more complicated in this case

form of differential equations and variational approaches. Figure 3.4 illustrates this situation, where, as mentioned before, all categories on the left use formalisation functor S_1, while all categories on the right use formalisation functor $S_2 = \vartheta_1(S_1)$.

Finally, let us discuss another alternative to build a global model from categories $\mathbf{Model}_i$, $i = 1, \ldots, 5$. Figure 3.2 presented an approach, where only one category has been coupled with another category on each step. Evidently, it is also possible to couple multiple categories simultaneously. Figure 3.5 illustrates one of many possibilities to perform such a coupling.

As it can be clearly seen from Fig. 3.5, the approach of coupling multiple categories simultaneously requires careful attention to the indices, which indicate the order in which the categories are coupled. In this regard, on the level of models, objects of

$$\mathbf{ModCoup}_{[(1,2),(2,3)],[(2,3),(3,4)],[(2,3),(3,4)],[(3,4),(4,5)]}$$

are formalised in form of the same expressions as objects of $\mathbf{ModCoup}_{1,2,3,4,5}$, but the structure of these objects is completely different and requires much more careful consideration to work with. Therefore, the approach to coupling several categories of mathematical models presented in Fig. 3.2 seems to be more transparent and provides a better internal structure for the respective categories of coupled mathematical models. Furthermore, in some applications, the order of model coupling might be predetermined from the beginning. For instance, geometrical restrictions can fix the order when constructing a global model of a building by sequentially coupling models for individual storeys.

Before providing an illustrative example, let us make few remarks:

Remark 3.17 All the constructions introduced in the previous subsections related to the complexity and convertibility of coupled mathematical models can also be extended to the case of coupling mathematical models from multiple categories. While this scaling is conceptually straightforward and can be realised inductively, it becomes very technical and will therefore not be presented here.

Remark 3.18 From the practical point of view, it is important to underline that often it is not required to keep track of all formal constructions presented above while constructing coupled models. In particular, after coupling two models, the resulting coupled models can be considered as one mathematical model and form a "simple" category of coupled mathematical models according to Definition 3.5:

- objects of the form $\langle \mathbf{Set}_i \cup \mathbf{Set}_j, \mathbf{Set}_{ij}^{\mathrm{Coup}} \rangle$ or $\langle \mathbf{Set}_i + \mathbf{Set}_i, \mathbf{Set}_{ii}^{\mathrm{Coup}} \rangle$ are converted into the sets of assumptions in the sense of Definition 3.1;
- the resulting category of mathematical models will be a partially ordered category in most cases, because of omitted concepts of base and coupling complexities.

Subsequently, these "simplified" coupled mathematical models can be coupled with mathematical model from other categories. For example, if the goal is to creating a global model of a structure by coupling simple one-dimensional beam models, then the coupling can be realised in a step-by-step manner. At each step, the models that are coupled should then be considered as a single mathematical model.

On the one hand, this approach simplifies the formal construction by reducing the number of indices and avoiding some of complicated abstract structures. But on the other hand, some care must still be taken by writing sets of assumptions for coupled models, particularly when models from the same category are being coupled. Otherwise, inconsistencies in the modelling process might appear.

Remark 3.19 Finally, let us remark, that the results presented in this subsection can also be viewed from the computer science perspective, particularly regarding a potential computer implementation of the category theory-based modelling framework. To implement the coupling of models from multiple categories, it is essential to know the formal structure of the objects of the resulting category of coupled mathematical models. Without this understanding, the correctness of the implementation cannot be verified. Although the topic of computer implementation of the category theory-based modelling framework goes beyond the scope of this book, we will provide a few remarks on it in Chap. 5.

Let us now consider an illustrative example of coupling models from three categories. We begin once more with a heat conduction process in a domain made of two different materials:

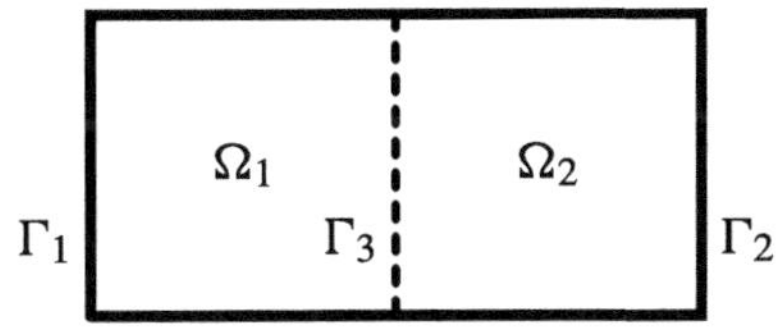

This example leads to the category **HeatCoup** with a coproduct structure. Next, we introduce the third category, **Elast**, and we thus interested in modelling thermoelasticity in a domain made of two materials with different heat conduction properties, but the same elasticity constants. We denote this resulting category of coupled mathematical models as **ThermoElastTwo**, which formally has the product structure

$$\mathbf{ThermoElastTwo} = \mathbf{Elast} \times (\mathbf{Heat} + \mathbf{Heat})\,.$$

For illustrative purposes, we will only show a few objects from this category, as the construction is straightforward. However, it is interesting to observe how the order in which categories are coupled influences the formal structure of the objects. Therefore, let us only consider the object $\mathbf{Set}_{\mathrm{LE}}$ of **Elast**, which contains assumptions of linear elasticity, and explore some of its couplings with objects from **Heat**. The objects of **ThermoElastTwo** have structure described in Definition 3.14. Thus, for $\mathbf{Set}_{\mathrm{LE}}$ we have

$$\begin{aligned}
\mathbf{Set}^{\mathrm{CDT}}_{\mathrm{LE,UH,UH}} &:= \langle \mathbf{Set}_{\mathrm{LE}}, \mathbf{Set}_{\mathrm{UH}} + \mathbf{Set}_{\mathrm{UH}}, \mathbf{Set}^{\mathrm{Coup}}_{\mathrm{UH,UH}} \cup \mathbf{Set}^{\mathrm{Coup}}_{\mathrm{CDT}} \rangle,\\
\mathbf{Set}^{\mathrm{CQST}}_{\mathrm{LE,UH,UH}} &:= \langle \mathbf{Set}_{\mathrm{LE}}, \mathbf{Set}_{\mathrm{UH}} + \mathbf{Set}_{\mathrm{UH}}, \mathbf{Set}^{\mathrm{Coup}}_{\mathrm{UH,UH}} \cup \mathbf{Set}^{\mathrm{Coup}}_{\mathrm{CQST}} \rangle,\\
\mathbf{Set}^{\mathrm{DT}}_{\mathrm{LE,UH,UH}} &:= \langle \mathbf{Set}_{\mathrm{LE}}, \mathbf{Set}_{\mathrm{UH}} + \mathbf{Set}_{\mathrm{UH}}, \mathbf{Set}^{\mathrm{Coup}}_{\mathrm{UH,UH}} \cup \mathbf{Set}^{\mathrm{Coup}}_{\mathrm{DT}} \rangle,
\end{aligned}$$

$$\begin{aligned}
\mathbf{Set}^{\mathrm{CDT}}_{\mathrm{LE,UH,NUH}} &:= \langle \mathbf{Set}_{\mathrm{LE}}, \mathbf{Set}_{\mathrm{UH}} + \mathbf{Set}_{\mathrm{NUH}}, \mathbf{Set}^{\mathrm{Coup}}_{\mathrm{UH,NUH}} \cup \mathbf{Set}^{\mathrm{Coup}}_{\mathrm{CDT}} \rangle,\\
\mathbf{Set}^{\mathrm{CQST}}_{\mathrm{LE,UH,NUH}} &:= \langle \mathbf{Set}_{\mathrm{LE}}, \mathbf{Set}_{\mathrm{UH}} + \mathbf{Set}_{\mathrm{NUH}}, \mathbf{Set}^{\mathrm{Coup}}_{\mathrm{UH,NUH}} \cup \mathbf{Set}^{\mathrm{Coup}}_{\mathrm{CQST}} \rangle,\\
\mathbf{Set}^{\mathrm{DT}}_{\mathrm{LE,UH,NUH}} &:= \langle \mathbf{Set}_{\mathrm{LE}}, \mathbf{Set}_{\mathrm{UH}} + \mathbf{Set}_{\mathrm{NUH}}, \mathbf{Set}^{\mathrm{Coup}}_{\mathrm{UH,NUH}} \cup \mathbf{Set}^{\mathrm{Coup}}_{\mathrm{DT}} \rangle,
\end{aligned}$$

$$\begin{aligned}
\mathbf{Set}^{\mathrm{CDT}}_{\mathrm{LE,UH,RH}} &:= \langle \mathbf{Set}_{\mathrm{LE}}, \mathbf{Set}_{\mathrm{UH}} + \mathbf{Set}_{\mathrm{RH}}, \mathbf{Set}^{\mathrm{Coup}}_{\mathrm{UH,RH}} \cup \mathbf{Set}^{\mathrm{Coup}}_{\mathrm{CDT}} \rangle,\\
\mathbf{Set}^{\mathrm{CQST}}_{\mathrm{LE,UH,RH}} &:= \langle \mathbf{Set}_{\mathrm{LE}}, \mathbf{Set}_{\mathrm{UH}} + \mathbf{Set}_{\mathrm{RH}}, \mathbf{Set}^{\mathrm{Coup}}_{\mathrm{UH,RH}} \cup \mathbf{Set}^{\mathrm{Coup}}_{\mathrm{CQST}} \rangle,\\
\mathbf{Set}^{\mathrm{DT}}_{\mathrm{LE,UH,RH}} &:= \langle \mathbf{Set}_{\mathrm{LE}}, \mathbf{Set}_{\mathrm{UH}} + \mathbf{Set}_{\mathrm{RH}}, \mathbf{Set}^{\mathrm{Coup}}_{\mathrm{UH,RH}} \cup \mathbf{Set}^{\mathrm{Coup}}_{\mathrm{DT}} \rangle,
\end{aligned}$$

$$\begin{aligned}
\mathbf{Set}^{\mathrm{CDT}}_{\mathrm{LE,UH,NURH}} &:= \langle \mathbf{Set}_{\mathrm{LE}}, \mathbf{Set}_{\mathrm{UH}} + \mathbf{Set}_{\mathrm{NURH}}, \mathbf{Set}^{\mathrm{Coup}}_{\mathrm{UH,NURH}} \cup \mathbf{Set}^{\mathrm{Coup}}_{\mathrm{CDT}} \rangle,\\
\mathbf{Set}^{\mathrm{CQST}}_{\mathrm{LE,UH,NURH}} &:= \langle \mathbf{Set}_{\mathrm{LE}}, \mathbf{Set}_{\mathrm{UH}} + \mathbf{Set}_{\mathrm{NURH}}, \mathbf{Set}^{\mathrm{Coup}}_{\mathrm{UH,NURH}} \cup \mathbf{Set}^{\mathrm{Coup}}_{\mathrm{CQST}} \rangle,\\
\mathbf{Set}^{\mathrm{DT}}_{\mathrm{LE,UH,NURH}} &:= \langle \mathbf{Set}_{\mathrm{LE}}, \mathbf{Set}_{\mathrm{UH}} + \mathbf{Set}_{\mathrm{NURH}}, \mathbf{Set}^{\mathrm{Coup}}_{\mathrm{UH,NURH}} \cup \mathbf{Set}^{\mathrm{Coup}}_{\mathrm{DT}} \rangle.
\end{aligned}$$

Continuing this process, we will obtain all possible coupled models. However, if we change the order in which models were coupled, then we would end with the following structure

$$\mathbf{ThermoElastTwo} = (\mathbf{Heat} \times \mathbf{Elast}) \times \mathbf{Heat},$$

but the objects would have the following structure

$$\mathbf{Set}^{\mathrm{CDT}}_{\mathrm{UH,LE,UH}} := \langle (\mathbf{Set}_{\mathrm{UH}} \cup \mathbf{Set}_{\mathrm{LE}}) + \mathbf{Set}_{\mathrm{UH}}, \mathbf{Set}^{\mathrm{Coup}}_{\mathrm{UH,UH}} \cup \mathbf{Set}^{\mathrm{Coup}}_{\mathrm{CDT}} \rangle,$$

where the disjoint union of sets of assumptions is necessary for keeping track that models from the same category are being coupled.

As clearly shown in the above examples, the internal structure of the objects of the categories of coupled mathematical models is different. However, when considering the level of model formulations, both approaches to constructing **ThermoElastTwo** lead to the same model formulations. It follows immediately from the fact that the same sets of assumptions are used in both cases. This observation further emphasises the importance of distinguishing between modelling assumptions and model formulations.

3.4 More Abstract Constructions on Categories of Mathematical Models

In this section, we provide some more abstract constructions on categories of mathematical models, as well as on categories of coupled mathematical models. We begin with a few remarks on functor categories, followed by a discussion of comma categories, and we will see, that they naturally arise during the construction of coupled mathematical models. Additionally, we will provide the meaning to a universal mapping property in the context of mathematical models. Finally, we will discuss an alternative perspective on mathematical models, which streamlines some of the constructions introduced in this chapter.

3.4.1 Functor Categories

In this subsection, we briefly introduce functor categories, which have implicitly appeared in the constructions from previous sections. We have discussed formalisation mappings of basic sets of assumptions and noted that these mappings are functorial. Therefore, it makes sense to fill the gap and formally define the following category:

Definition 3.16 For categories of mathematical models **Model**, and categories of coupled mathematical models $\mathbf{ModCoup}_{i,j}$ and $\mathbf{ModCoup}_{i,i}$, the categories of the corresponding formalised sets of basic assumptions will be denoted by $\mathbf{Model}^F$, $\mathbf{ModCoup}^F_{i,j}$, and $\mathbf{ModCoup}^F_{i,i}$, respectively.

Now we list the functor categories that have already implicitly appeared:

- Functor categories of formalisation mappings:

$$\begin{aligned} \left(\mathbf{Model}^F\right)^{\mathbf{Model}} &= \mathrm{Funct}\left(\mathbf{Model}, \mathbf{Model}^F\right), \\ \left(\mathbf{ModCoup}^F_{i,j}\right)^{\mathbf{ModCoup}_{i,j}} &= \mathrm{Funct}\left(\mathbf{ModCoup}_{i,j}, \mathbf{ModCoup}^F_{i,j}\right); \end{aligned}$$

- Functor categories related to coupled models:

$$\begin{aligned} \left(\mathbf{ModCoup}_{i,j}\right)^{\mathbf{Model}_i} &= \mathrm{Funct}\left(\mathbf{Model}_i, \mathbf{ModCoup}_{i,j}\right), \\ \left(\mathbf{ModCoup}_{i,j}\right)^{\mathbf{Model}_j} &= \mathrm{Funct}\left(\mathbf{Model}_j, \mathbf{ModCoup}_{i,j}\right). \end{aligned}$$

3.4.2 Comma Categories

In fact, several comma categories have already arisen implicitly during the construction of the category theory-based modelling framework. We now make them explicit by discussing several cases related to Definition 2.11. We start with the simplest case and the categories of mathematical models **Model**.

Let $\mathbf{Set}_1$ be an object of the category **Model**, then the category of objects under $\mathbf{Set}_1$ is the category $(\mathbf{Set}_1 \downarrow \mathbf{Model})$ for which holds:

- objects are all pairs $\langle f, \mathbf{Set}_i \rangle$, $i \neq 1$, where $\mathbf{Set}_i$ is an object of **Model** and $f\colon \mathbf{Set}_1 \longrightarrow \mathbf{Set}_i$ an arrow of **Model**;
- arrows are $h\colon \langle f, \mathbf{Set}_i \rangle \longrightarrow \langle f', \mathbf{Set}_i' \rangle$ those arrows $h\colon \mathbf{Set}_i \longrightarrow \mathbf{Set}_i'$ of **Model** for which $h \circ f = f'$.

This definition is represented in a diagrammatic form as follows:

$$\text{objects } \langle f, \mathbf{Set}_i \rangle : \quad \begin{array}{c} \mathbf{Set}_1 \\ \big\downarrow f \\ \mathbf{Set}_i \end{array}$$

$$\text{arrows } \langle f, \mathbf{Set}_i \rangle \xrightarrow{h} \langle f', \mathbf{Set}_i' \rangle : \quad \begin{array}{ccc} & \mathbf{Set}_1 & \\ {\scriptstyle f}\swarrow & & \searrow{\scriptstyle f'} \\ \mathbf{Set}_i & \xrightarrow{h} & \mathbf{Set}_i' \end{array}$$

The consideration of comma categories of the form $(\mathbf{Set}_1 \downarrow \mathbf{Model})$ for mathematical models provides a possibility to further classify models. For example, by selecting modelling assumptions of particular interest, a comma category can be created to explore which sets of assumptions can be derived from it and which morphisms are necessary for that. This additional structural information enhances a better understanding of models comparability, particularly for practical applications, see for example [13], as well as discussion in Chap. 4.

The construction of comma category presented above can be straightforwardly extended to the case of coupled models. For the case of coupling of models from different categories, the category of objects under $\mathbf{Set}_{A,B}$ is the category $\left(\mathbf{Set}_{A,B} \downarrow \mathbf{ModCoup}_{i,j}\right)$ with the following diagrammatic representation:

$$\text{objects } \langle f, \mathbf{Set}_{i,j} \rangle : \quad \begin{array}{c} \mathbf{Set}_{A,B} \\ \Big\downarrow {\scriptstyle f} \\ \mathbf{Set}_{i,j} \end{array}$$

$$\text{arrows } \langle f, \mathbf{Set}_{i,j} \rangle \xrightarrow{h} \langle f', \mathbf{Set}_{i,j}' \rangle : \quad \begin{array}{ccc} & \mathbf{Set}_{A,B} & \\ {\scriptstyle f}\swarrow & & \searrow{\scriptstyle f'} \\ \mathbf{Set}_{i,j} & \xrightarrow{h} & \mathbf{Set}_{i,j}' \end{array}$$

Similarly, the following diagram represents the structure of the category of objects under $\mathbf{Set}_{A,A}$, i.e. $\left(\mathbf{Set}_{A,A} \downarrow \mathbf{ModCoup}_{i,i}\right)$, for the case of coupling of models from the same category:

$$\text{objects } \langle f, \mathbf{Set}_{i,i} \rangle : \quad \begin{array}{c} \mathbf{Set}_{A,A} \\ \Big\downarrow {\scriptstyle f} \\ \mathbf{Set}_{i,i} \end{array}$$

arrows $\langle f, \mathbf{Set}_{i,i} \rangle \xrightarrow{h} \langle f', \mathbf{Set}'_{i,i} \rangle$:

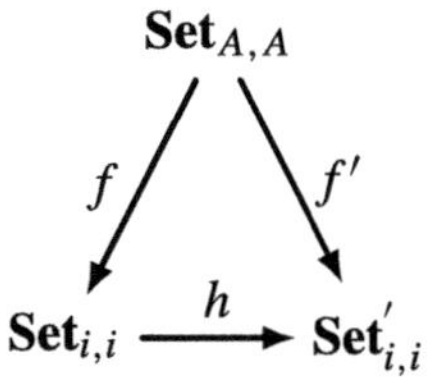

The practical interpretation of these comma categories is analogous to that of categories of mathematical models discussed above.

Analogously, we can define the following comma categories:

- Let $\mathbf{Set}_1$ be an object of the category **Model**, then the category of objects over $\mathbf{Set}_1$ is the category ($\mathbf{Model} \downarrow \mathbf{Set}_1$) with the following diagram:

- Let $\mathbf{Set}_{A,B}$ be an object of the category $\mathbf{ModCoup}_{i,j}$, then the category of objects over $\mathbf{Set}_{A,B}$ is the category $\left(\mathbf{ModCoup}_{i,j} \downarrow \mathbf{Set}_{A,B}\right)$ with the following diagram:

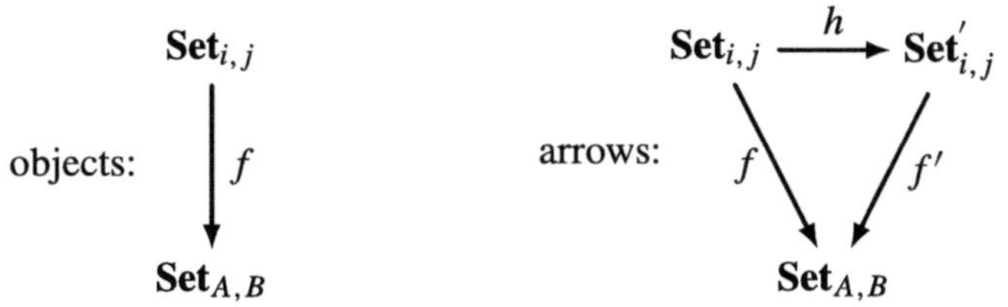

- Let $\mathbf{Set}_{A,A}$ be an object of the category $\mathbf{ModCoup}_{i,i}$, then the category of objects over $\mathbf{Set}_{A,A}$ is the category $\left(\mathbf{ModCoup}_{i,i} \downarrow \mathbf{Set}_{A,A}\right)$ with the following diagram:

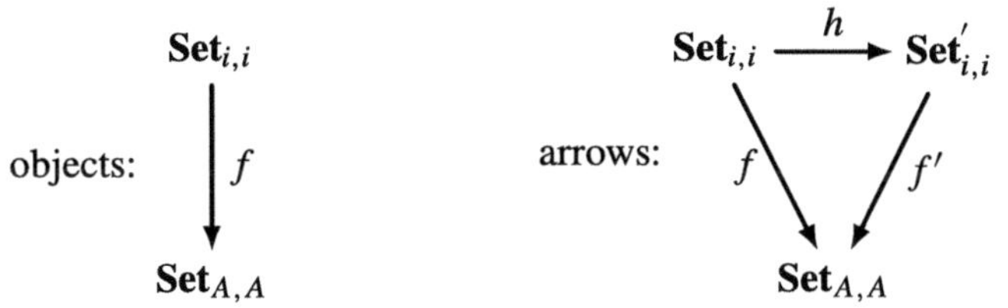

As in the case of the comma categories $(\mathbf{Set}_1 \downarrow \mathbf{Model})$, the categories $(\mathbf{Model} \downarrow \mathbf{Set}_1)$ of objects over $\mathbf{Set}_1$ provide a possibility to further classify models, by studying which sets of assumptions lead to a mathematical model with modelling assumptions of particular interest and which morphisms are used on the way.

Next, we consider the case then functors are involved in the construction of comma categories. To illustrate this case, we begin once more with the categories of mathematical models. Let A_1 be an object of the category $\mathbf{Model}^F$ and $S\colon \mathbf{Model} \to \mathbf{Model}^F$ be a formalisation functor, then the category of objects S-over A_1 is the category $(S \downarrow \mathrm{A}_1)$ for which holds:

- objects are all pairs $\langle f, \mathbf{Set}_i \rangle$ with $\mathbf{Set}_i$ being an object of $\mathbf{Model}$ and $f\colon S(\mathbf{Set}_i) \longrightarrow A_1$;
- arrows are $h\colon \langle f, \mathbf{Set}_i \rangle \longrightarrow \langle f', \mathbf{Set}_i^{'} \rangle$ all those arrows $h\colon \mathbf{Set}_i \longrightarrow \mathbf{Set}_i^{'}$ in $\mathbf{Model}$ for which $f' = Sh \circ f$.

The following diagram represents this definition:

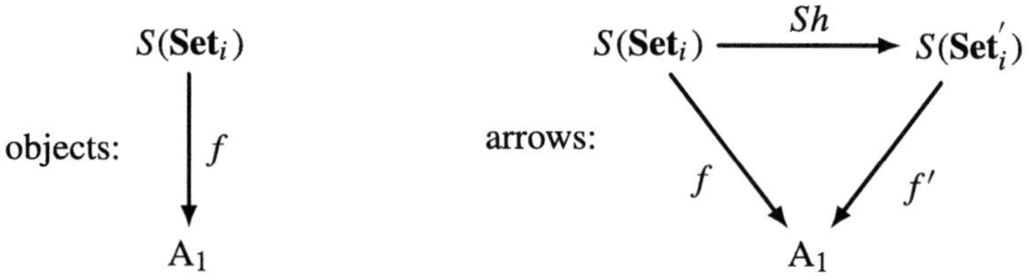

The practical meaning of this definition is as follows: as discussed previously, there exist different ways of formalising the sets of assumptions of mathematical models. It is essential to establish a connection between these formalisation ways and the final representation of a mathematical model. The comma category $(S \downarrow \mathrm{A}_1)$ serves to provide this connection.

Let us make the following remark:

Remark 3.20 Formally, it is also possible to introduce the comma category $(\mathrm{A}_1 \downarrow T)$ of the objects T-under A_1. However, the practical significance of this definition would be rather limited. As discussed in the context of Definition 3.1, a formalisation functor S is generally not invertible, which is a necessary condition for constructing the functor T. Therefore, we omit the definition of $(\mathrm{A}_1 \downarrow T)$ here.

Since coupled models are also formalised by the help of functorial mappings, the corresponding comma categories can be constructed in a completely analogous manner and therefore omitted here. However, the categories of coupled mathematical models give rise to another type of comma categories, which we will discuss next. Let $\mathbf{Set}_{1,j}$ be an object of the category $\mathbf{ModCoup}_{1,j}$ and $F\colon \mathbf{Model}_1 \to \mathbf{ModCoup}_{1,j}$, then the category of objects F-under $\mathbf{Set}_{1,j}$ is the category $\left(\mathbf{Set}_{1,j} \downarrow F\right)$ for which holds:

- objects are all pairs $\langle f, \mathbf{Set}_i \rangle$ with $\mathbf{Set}_i$ being an object of $\mathbf{Model}_1$ and $f\colon \mathbf{Set}_{1,j} \longrightarrow F(\mathbf{Set}_i)$;

- arrows are $h\colon \langle f, \mathbf{Set}_i\rangle \longrightarrow \langle f', \mathbf{Set}'_i\rangle$ all those arrows $h\colon \mathbf{Set}_i \longrightarrow \mathbf{Set}'_i$ in $\mathbf{Model}_1$ for which $f' = Sh \circ f$.

The following diagram represents this definition:

$$\text{objects:}\quad \begin{array}{c} \mathbf{Set}_{i,j} \\ \downarrow{\scriptstyle f} \\ F(\mathbf{Set}_i) \end{array} \qquad\qquad \text{arrows:}\quad \begin{array}{ccc} & \mathbf{Set}_{i,j} & \\ {\scriptstyle f}\swarrow & & \searrow{\scriptstyle f'} \\ F(\mathbf{Set}_i) & \xrightarrow{\quad Sh \quad} & F(\mathbf{Set}'_i) \end{array}$$

The practical meaning of this comma category lies in analysing which sets of basic assumption of specific mathematical models on the left contribute to construction of the set of assumptions $\mathbf{Set}_{1,j}$ of a coupled mathematical model. In fact, this comma category has been already implicitly discussed during construction of category **ThermoElast** in the previous section. Additionally, if now $\mathbf{Set}_2 \in \mathbf{Model}_2$ and $T\colon \mathbf{Model}_2 \to \mathbf{ModCoup}_{i,2}$, then the category $(T \downarrow \mathbf{Set}_2)$ of objects F-over $\mathbf{Set}_2$ can be constructed. This basically means, that we consider now sets of assumptions on the right in the respective category of coupled mathematical models. The both situations are covered by the general case:

Definition 3.17 Let $\mathbf{Model}_i$ and $\mathbf{Model}_j$ with arbitrary i, j be two categories of mathematical models, and let $\mathbf{ModCoup}_{i,j}$ denotes the category of coupled models constructed by help of two functors

$$\mathbf{Model}_i \stackrel{T}{\longrightarrow} \mathbf{ModCoup}_{i,j} \stackrel{S}{\longleftarrow} \mathbf{Model}_j.$$

The *comma category* $(T \downarrow S)$, also written (T, S), has as objects all triples $\langle \mathbf{Set}_i, \mathbf{Set}_j, f\rangle$, with $\mathbf{Set}_j$ being an object of $\mathbf{Model}_j$, $\mathbf{Set}_i$ being an object of $\mathbf{Model}_i$, and $f\colon T(\mathbf{Set}_i) \longrightarrow S(\mathbf{Set}_j)$, and as arrows $\langle \mathbf{Set}_i, \mathbf{Set}_j, f\rangle \longrightarrow \langle \mathbf{Set}'_i, \mathbf{Set}'_j, f'\rangle$ all pairs $\langle k, h\rangle$ of arrows $k\colon \mathbf{Set}_i \longrightarrow \mathbf{Set}'_i$, $h\colon \mathbf{Set}_j \longrightarrow \mathbf{Set}'_j$ such that $f' \circ Tk = Sh \circ f$. In diagrams,

$$\text{objects } \langle e, d, f\rangle\text{:}\quad \begin{array}{c} T(\mathbf{Set}_i) \\ \downarrow{\scriptstyle f} \\ S(\mathbf{Set}_j) \end{array} \qquad \text{arrows } \langle k, h\rangle\text{:}\quad \begin{array}{ccc} T(\mathbf{Set}_i) & \xrightarrow{\quad Tk \quad} & T(\mathbf{Set}'_i) \\ {\scriptstyle f}\downarrow & & \downarrow{\scriptstyle f'} \\ S(\mathbf{Set}_j) & \xrightarrow{\quad Sh \quad} & S(\mathbf{Set}'_i) \end{array}$$

with the square commutative. The composite $\langle k', h'\rangle \circ \langle k, h\rangle$ is $\langle k' \circ k, h' \circ h\rangle$, when defined.

Again, the practical interpretation of this construction is that we can split the category of coupled mathematical models $\mathbf{ModCoup}_{i,j}$ into classes related to specific sets of assumptions from categories $\mathbf{Model}_i$ and $\mathbf{Model}_j$ of particular interest, and analyse which coupled models of different complexity can be constructed.

3.4.3 Universal Arrows

Similar to the comma categories discussed in the previous subsection, there are several interpretations of universal arrows in the context of category theory-based modelling framework. In this section, we will discuss these possible interpretations.

Let us consider a formalisation functor $S\colon \mathbf{Model} \to \mathbf{Model}^F$, where $\mathbf{Model}^F$ is a category of formalised sets of assumptions, or, in other words, a category of mathematical models formulations corresponding to the objects in **Model**. For keeping notations consistent, let us denote objects of **Model** by $\mathbf{Set}_i$, $i = 1, 2, \ldots$ and objects of $\mathbf{Model}^F$ by $S\,\mathbf{Set}_i$. Then a universal arrow from $S\,\mathbf{Set}_i$ to S is a pair $\langle \mathbf{Set}_j, u\rangle$ consisting of an object $\mathbf{Set}_j$ of **Model** and an arrow $u\colon S\,\mathbf{Set}_i \to S\,\mathbf{Set}_j$ of $\mathbf{Model}^F$, such that to every pair $\langle \mathbf{Set}_k, f\rangle$ with $\mathbf{Set}_k$ an object of **Model** and $f\colon S\,\mathbf{Set}_i \to S\,\mathbf{Set}_k$ an arrow of $\mathbf{Model}^F$, there is a unique arrow $f'\colon \mathbf{Set}_j \to \mathbf{Set}_k$ of **Model** with $Sf' \circ u = f$. This construction corresponds to the following diagram, see also Definition 2.12:

$$\begin{array}{ccccc} S\,\mathbf{Set}_i & \xrightarrow{\ u\ } & S\,\mathbf{Set}_j & \qquad & \mathbf{Set}_j \\ \| & & \Big\downarrow{\scriptstyle Sf'} & & \Big\downarrow{\scriptstyle f'} \\ S\,\mathbf{Set}_i & \xrightarrow{\ f\ } & S\,\mathbf{Set}_k & & \mathbf{Set}_k \end{array}$$

Let us now discuss this construction of a universal arrow on the example of the category **Beam**, which has three objects $\mathbf{Set}_{\mathrm{B-E}}$, $\mathbf{Set}_{\mathrm{R}}$, and $\mathbf{Set}_{\mathrm{T}}$, containing the sets of assumptions of different beam theories. In this case, we have the following diagram for the universal arrow in **Beam**:

$$\begin{array}{ccccc} S\,\mathbf{Set}_{\mathrm{B-E}} & \xrightarrow{\ Sf\ } & S\,\mathbf{Set}_{\mathrm{R}} & \qquad & \mathbf{Set}_{\mathrm{R}} \\ \| & & \Big\downarrow{\scriptstyle Sg} & & \Big\downarrow{\scriptstyle g} \\ S\,\mathbf{Set}_{\mathrm{B-E}} & \xrightarrow{\ Sh\ } & S\,\mathbf{Set}_{\mathrm{T}} & & \mathbf{Set}_{\mathrm{T}} \end{array}$$

Thus, the universal arrow from $S\,\mathbf{Set}_{\mathrm{B-E}}$ to S is a pair $\langle \mathbf{Set}_{\mathrm{R}}, Sf\rangle$. Note that we have adapted here notations for morphisms in $\mathbf{Beam}^F$ used in Sect. 3.2.4.1.

Let us now discuss several aspects of universal arrows from the modelling perspective:

- Universal arrows link categories of mathematical models, whose objects are sets of assumptions, with categories of model formulations, where the objects are various (equivalent) model formulations. A possible interpretation of this is that if a category **Model** is a totally ordered, then the corresponding category $\mathbf{Model}^F$ can also be made totally ordered, even if different formalisms are used to create specific model formulations.
- The completion to a totally ordered category is achieved through a universal arrow, involving the respective arrows u and f'. This process may also involve natural transformations of the formalisation functor S, enabling translations between different formalisations when necessary.
- Further, we can conclude, that if two sets of assumptions are related (via complexity), then their model formulations must be related as well, even if they are formalised by different formalisation approaches.

In summary, the universal arrows can be interpreted as *order-enforcing mapping for model formulations*. Consequently, we can formulate the following corollary:

Corollary 3.6 *A universal arrow relating* **Model** *and* $\mathbf{Model}^F$ *exist, if* **Model** *is a totally ordered category.*

In the case of a partially ordered category, the condition that for every pair to every pair $\langle \mathbf{Set}_k, f \rangle$ there is a unique arrow $f' \colon \mathbf{Set}_j \to \mathbf{Set}_k$ will fail. However, it is always possible to extract a totally ordered subcategory from a partially ordered category, if necessary.

The discussion about universal arrows presented above can be straightforwardly uplifted to the case of coupled mathematical models. The corresponding interpretations will be the similar to the case of categories of mathematical models.

Further, there is another perspective on universal arrows in the context of coupled mathematical models. Let $\mathbf{ModCoup}_{i,j}$ denote the category of coupled mathematical models with arrows being relations discussed in Sect. 3.3.2, and let $F \colon \mathbf{ModCoup}_{i,j} \to \mathbf{Model}_i$ be a forgetful functor sending each set of assumptions of coupled mathematical models to the set of assumptions $\mathbf{Set}_i$ on the left, i.e. to an object of $\mathbf{Model}_i$. Then for any set of assumptions $\mathbf{Set}_i$, there is a corresponding set of assumptions of coupled mathematical models $\mathbf{Set}_{i,j}$. The function which sends each object $\mathbf{Set}_i$ of $\mathbf{Model}_i$ to the same object interpreted as a subset of $\mathbf{Set}_{i,j}$ is an arrow $j \colon \mathbf{Set}_i \to F\left(\mathbf{Set}_{i,j}\right)$. Then for any other object of $\mathbf{ModCoup}_{i,j}$, e.g. $\mathbf{Set}'_{i,j}$, each function $f \colon \mathbf{Set}_i \to F\left(\mathbf{Set}'_{i,j}\right)$ can be uniquely extended to a relation between coupled mathematical models $f' \colon \mathbf{Set}_{i,j} \to \mathbf{Set}'_{i,j}$ with $Ff' \circ j = f$, meaning that j is a universal arrow from $\mathbf{Set}_i$ to F. We have now the following diagram:

$$
\begin{array}{ccccc}
\mathbf{Set}_i & \xrightarrow{\ j\ } & F\left(\mathbf{Set}_{i,j}\right) & \qquad & \mathbf{Set}_{i,j} \\
\| & & \Big\downarrow{\scriptstyle Ff'} & & \Big\downarrow{\scriptstyle f'} \\
\mathbf{Set}_i & \xrightarrow{\ f\ } & F\left(\mathbf{Set}'_{i,j}\right) & & \mathbf{Set}'_{i,j}
\end{array}
$$

The practical interpretations of this construction are similar to the one discussed above for categories of mathematical models and their formalisations:

- If coupled mathematical models are base complexity related (left or right), then evidently the respective "individual" mathematical models are also complexity related.
- If two coupled mathematical models are solely related in terms of coupling complexity, then $j = f$ and consequently $f' = \mathrm{id}$. This implies that the construction of a universal arrow, as described above, is not applicable for such cases. However, if we focus on the sets of assumptions and their formalisations, then it is possible to apply the interpretations of universal arrows discussed previously.
- An analogous construction is also possible for the case of coupling of models from the same category, i.e. for categories $\mathbf{ModCoup}_{i,i}$. However, this result is trivial, since ordering of objects in $\mathbf{ModCoup}_{i,i}$ with respect to the base complexity is identical to ordering of objects in $\mathbf{Model}_i$, and, therefore, only coupling complexity is different.

Finally, for making the statement about a unique relation between coupled mathematical models $f'\colon \mathbf{Set}_{i,j} \to \mathbf{Set}'_{i,j}$ hold, it is necessary to consider only totally ordered categories $\mathbf{Model}_i$ and $\mathbf{Model}_j$. Again, we summarise it in the form of the following corollary:

Corollary 3.7 *Let* $\mathbf{ModCoup}_{i,j}$ *be a category of coupled mathematical models coupling categories* $\mathbf{Model}_i$ *and* $\mathbf{Model}_j$*. Then the following statements hold:*

- *A universal arrow relating* $\mathbf{Model}_i$ *and* $\mathbf{ModCoup}_{i,j}$ *exist, if* $\mathbf{Model}_i$ *is a totally ordered category.*
- *A universal arrow relating* $\mathbf{Model}_j$ *and* $\mathbf{ModCoup}_{i,j}$ *exist, if* $\mathbf{Model}_j$ *is a totally ordered category.*

Otherwise, universal arrows can be constructed only for totally ordered subcategories of $\mathbf{Model}_i$ *and* $\mathbf{Model}_j$*.*

3.5 Summary of the Chapter

In this chapter, we have introduced the category theory-based modelling framework to provide a structured perspective and a better understanding of mathematical models. In this framework, mathematical models are not described directly, but rather associated with basic sets of modelling assumptions from which these models are derived. The separation between the models themselves and the underlying sets of assumptions is critical for a consistent categorical description of mathematical models. In this way, the notion of model complexity, which relates different mathematical models, can be defined meaningfully.

As it is well known from the engineering practice, the same set of assumptions can often be formalised by using different mathematical tools leading to different model formulations, as for example a differential form and a variational form of the same problem. This fact is addressed in the category theory-based modelling framework through formalisation functors and their natural transformations, which leads to the notion of convertible mathematical models.

After providing a categorical description of mathematical models, coupled models are addressed. It is not surprising that the structure of objects in categories of coupled mathematical models is more complicated due to the additional coupling conditions and combinations of individual models. Two categorical structures are utilised to described coupled mathematical models:

- Product for coupling models from different categories, and its dual
- Coproduct for coupling models from the same category.

The notions of model complexity and convertibility are then extended to the case of coupled mathematical models. Additionally, the case of coupling of multiple categories of mathematical models has also been discussed, where the concept of the lattice of theories has been used for a schematic representation of the coupling process.

Finally, abstract constructions on categories of mathematical models, such as functor categories, comma categories, and universal arrows, have been discussed. These abstract constructions provide another perspectives on mathematical models and have been also interpreted from the modelling perspective.

It is necessary to remark, that the results presented in this chapter constitute the foundation of category theory-based modelling framework. Some of the categorical constructions presented are, in fact, quite technical, which is to be expected when working with formal mathematics. This technicality is not a drawback of the approach, but rather a necessary step for a formal verification of the construction of coupled mathematical models.

References

1. Awodey, S. (2010). *Category theory*. Oxford University Press Inc.
2. Bock, S. (2009). *Über funktionentheoretische Methoden in der räumlichen Elastizitätstheorie*. PhD Thesis, Bauhaus-Universität Weimar.
3. Bock, S., Gürlebeck, K., Legatiuk, D., & Nguyen, H. M. (2015). ψ-hyperholomorphic functions and a Kolosov-Muskhelishvili formula. *Mathematical Methods in the Applied Sciences, 38*(18), 5114–5123.
4. Eringen, A. C. (1966). Linear theory of micropolar elasticity. *Journal of Mathematics and Mechanics, 15*(6), 909–923.
5. Erofeev, V. I., Kazhaev, V. V., & Semerikova, N. P. (2002). *Waves in bars. Dispersion. Dissipation. Nonlinearity*. Fismatlit (in Russian).
6. Grisvard, P. (1992). *Singularities in boundary value problems*. Springer.
7. Gürlebeck, K., & Sprößig, W. (1997). *Quaternionic and Clifford calculus for physicists and engineers*. John Wiley & Sons.
8. Gürlebeck, K., Habetha, K., & Sprößig, W. (2016). *Application of holomorphic functions in two and higher dimensions*, Birkhäuser.
9. Gürlebeck, K., Hofmann, D., & Legatiuk, D. (2017). Categorical approach to modelling and to coupling of models. *Mathematical Methods in the Applied Sciences, 40*(3), 523–534.
10. Gürlebeck, K., & Legatiuk, D. (2019). Quaternionic operator calculus for boundary value problems of micropolar elasticity. In *Topics in Clifford analysis*. Trends in mathematics (pp. 221–234). Springer.
11. Gürlebeck, K., Legatiuk, D., & Webber, K. (2022). Operator calculus approach to comparison of elasticity models for modelling of masonry structures. *Mathematics, 10*(10), 1670.
12. Hetnarski, R. B., & Eslami, M. R. (2019). *Thermal stresses – Advanced theory and applications*. Springer Nature Switzerland.
13. Kavrakov, I., Legatiuk, D., Gürlebeck, K., & Morgenthal, G. (2019). A categorical perspective towards aerodynamic models for aeroelastic analyses of bridges. *Royal Society Open Science, 6*, 181848.
14. Kovalenko, A. (1969). *Thermoelasticity: Basic theory and applications*. Wolters-Noordhoff.
15. Legatiuk, D. (2015). *Evaluation of the coupling between an analytical and a numerical solution for boundary value problems with singularities*. PhD Thesis, Bauhaus-Universität Weimar.
16. Liebowitz, H. (1968). *Fracture, an advanced treatise. Volume II: Mathematical fundamentals*. Academic Press.
17. Liu, W., & Röckner, M. (2015). *Stochastic partial differential equations: An introduction*. Springer International Publishing Switzerland.
18. Lurie, A. I. (1990). *Nonlinear theory of elasticity*. Elsevier Science Publishers B.V.
19. Lurie, A. I. (2005). *Theory of elasticity*. Foundations of engineering mechanics. Springer.
20. Muskhelishvili, N. I. (1977). *Some basic problems of the mathematical theory of elasticity*. Springer Science+Business Media Dordrecht.
21. Sowa, J. (2000). *Knowledge representation: Logical, philosophical, and computational foundations*. Brooks/Cole.
22. Tikhonov, A. N., & Samarskii, A. A. (1963). *Equations of mathematical physics*. Dover Publications.
23. Timoshenko, S. P. (1972). *Course of elasticity theory (Kurs teorii uprugosti)*. Dumka (in Russian).

Chapter 4
Engineering Perspectives and Applications

Abstract This chapter begins by presenting simpler analogues of the categorical constructions proposed in the previous chapter within the category theory-based modelling framework. The goal here is to grasp the essential ideas of these constructions in simpler formulations motivated by the engineering perspective. These formulations remain accurate, but are free of too many formal constructions. Consequently, the practical applicability of the category theory-based modelling framework proposed in this book will be increased. After that, we discuss some further model properties, such as for example model uncertainty and model robustness, and evaluate them with respect to their potential for formalisation within the category theory-based modelling framework. Additionally, we explore a more abstract perspective on model comparison and model evaluation. Finally, the chapter concludes with an application of the category theory-based modelling framework to practical problems of bridge aerodynamics and a discussion of how the categorical modelling perspective can support the development of solution methods for coupled problems in practice.

4.1 Engineering Perspective on Categorical Constructions

The aim of this section is to enhance the practical applications of the category theory-based modelling framework by providing simpler analogues of the key categorical constructions introduced in Chap. 3. This simplification is indented to help engineers better understand mathematical models, especially coupled models, without being overwhelmed with technical details of the abstract framework. Furthermore, for a better readability of this section, we will provide all analogues in the form of separate definitions, although these definitions may not necessary really define particular objects. Additionally, where it is necessary, we will explore the connection between the formal definition and its simplified analogue in more details.

Figure 4.1 presents a general overview of the motivation for providing the engineering perspective on abstract categorical constructions. The informal versions of the definitions act as an interlayer between practical problems and the category

D. Legatiuk, *Mathematical Modelling by Help of Category Theory*,
Mathematical Engineering, https://doi.org/10.1007/978-3-031-84951-0_4

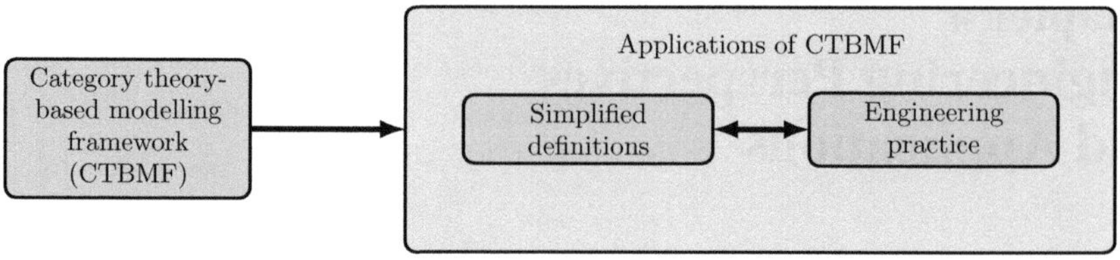

Fig. 4.1 Illustration of using the category theory-based modelling framework in practice: simplified version of the abstract definitions serve as an optional interlayer between the framework and practical problems

theory-based modelling framework. While this interlayer is purely optional and can be omitted if not needed, it provides flexibility for working with the abstract constructions without compromising correctness. In practical terms, this approach allows to explore a given problem at first informally for creating a general "picture" and discussing the types of models involved or required. After this informal stage, rigorous categorical constructions can be utilised to ensure correct descriptions, if necessary. We will revisit this idea of working with less formal constructions to enhance the applicability of the category theory-based modelling framework later in this chapter, as well as in Chap. 5.

At first, we provide an informal definition for categories of mathematical models introduced in Definition 3.1. This abstract definition can be reformulated from the practical point of view as follows:

Definition 4.1 (Mathematical Models in Engineering) A mathematical model is a combination of the basic modelling assumptions, the model formulation in terms of mathematical expression, and the derivation of the model formulation from the modelling assumptions. Further, the primary difference between different mathematical models lies in their modelling assumptions.

This definition brings together the discussion on Definition 3.1 and the notion of model complexity introduced in Definition 3.2. Specifically, the statement "mathematical models primarily differ in their modelling assumptions" is an informal way of describing the relations between sets of modelling assumptions in terms of inclusions. Furthermore, Definition 4.1 emphasises once more that a mathematical model is more than just a single equation or system of equations. It highlights that the differences between models lie in their underlying modelling assumptions, and not in their final formulations. This perspective on mathematical models will be further generalised in Chap. 5 to explore the relational algebra-based approach to describing models and their couplings.

Although Definition 4.1 implicitly includes the notion of model complexity, it is still beneficial to provide a practical interpretation of the model complexity introduced in Definition 3.2. Consequently, we propose the following definition:

Definition 4.2 (Model Complexity) The complexity of a mathematical model measures the range of physical phenomena that the model can describe. In this context, models can be compared and ordered if there is an overlap in the physical phenomena they can describe.

In Chap. 3, we have already mentioned that the notion of model complexity can be seen as an abstract analogue of *model fidelity*. The fidelity itself can be seen as the distance between a model and the system it represents [6]. The term "distance" suggests that there exists a way to quantify the difference between the model and reality. In contrast, the model complexity, as introduced in Definition 3.2 and revisit in Definition 4.2, addresses this "distance" at the level of modelling assumptions. For example, if we are interested in the modelling of a non-linear phenomenon, but we choose a linear model, then it is evident from the beginning, that the model fidelity cannot be high in this case. Thus, model complexity provides a general setting for the range of applications for a given model. However, it is still possible to quantify the influence of specific modelling assumptions on the fidelity of the resulting model. We will introduce more formal definitions for such quantifications in Sect. 4.2 and present a practical engineering example in Sect. 4.3.

Finally, the discussion on individual mathematical models concludes with a revisit of the notion of convertibility of mathematical models. Model convertibility can be characterised as follows:

Definition 4.3 (Model Convertibility) The convertibility of model formulations implies that these formulations correspond to the same set of modelling assumptions. Thus, while different model formulations can be created for the same set of modelling assumptions, the model complexity remains the same.

This simplified description emphasises that different model formulations do not fundamentally change the model fidelity in the sense discussed above; and that these formulations can be transformed into one another (by the help of a natural transformation). Considering that a model formulation is often understood as a model in the engineering practice, a typical misconception is then to claim that one model formulation is more complex than another—for example that a model expressed in terms of differential equations is more complex than one in a variational form, or vice versa. This is exactly the ground where the model complexity starts to be confused with computational complexity or the overall complexity of a solution procedure. However, for a deeper understanding of models and their practical use, it is critical to keep a clear separation between the level of models and the level of their applications, as discussed also in Chap. 1. Definition 4.3 ensures that this separation is kept also in engineering practice.

For coupled mathematical models, the following practical explanation can be provided:

Definition 4.4 (Coupled Mathematical Models in Engineering) A coupled mathematical model is not just a system of equations, but rather a combination of two (or more) mathematical models. Each of these models has its own set of basic modelling assumptions, a model formulation in terms of mathematical expression, and a derivation of this model formulation from the modelling assumptions.

Additionally, a coupled mathematical model must have a set of coupling conditions, since otherwise, a combination of models would be still decoupled.

Evidently, this definition omits formal constructions in terms of products and coproducts introduced in Definitions 3.6–3.7, which were necessary to distinguish between the coupling of models from the same category and from different categories. Nonetheless, Definition 4.4 captures the main essence of coupled models and thus suffices for a practical understanding of their nature: the coupling of models always requires a clearly formulated coupling conditions.

Additionally, it is beneficial to characterise the coupling of mathematical models in the following way:

Definition 4.5 (Coupling of Mathematical Models) The coupling of mathematical models is a mapping between different mathematical models that increases the overall model complexity.

This definition summarises an intuitive understanding of coupled models: coupled models are able to describe more physical phenomena, than the individual models being coupled, consequently, their complexity, in the sense of Definition 4.2, increases. Of course, Definition 4.5 intentionally oversimplifies the notion of model complexity for coupled mathematical models, which has been discussed in details in Sect. 3.3.2. More comprehensive details regarding the construction of coupled mathematical modes within the category theory-based modelling framework are necessary when working with these models. However, Definition 4.5 provides an initial, practical understanding of coupled models and the general idea concept of coupling.

Next, we will provide alternative views on abstract constructions on categories of mathematical models, which have been presented in Sect. 3.4. We start with alternative descriptions of functor categories:

Definition 4.6 (Different Model Derivations) Using different methods to derive model formulations from their sets of modelling assumptions respects the ordering of models based on their complexity. Moreover, one derivation can be converted or translated into another.

Definition 4.7 (Different Model Couplings) The coupling of models is realised thought a family of mappings, which might be vary for models describing the same physical process but different in their complexity.

These definitions capture the essence of formal constructions of functor categories presented in Definition 3.16.

The practical meaning of comma categories arising in the category theory-based modelling framework can be summarised as follows:

Definition 4.8 (Model Classes) By selecting specific sets of modelling assumptions, model classes can be constructed, indicating which simpler or more complex models can arise from these assumptions, or which coupled models can be constructed based on them.

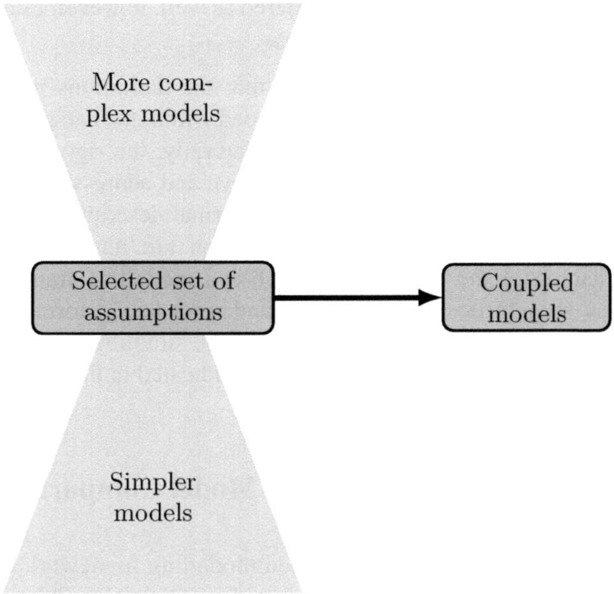

Fig. 4.2 Illustration of comma categories in the modelling context: a given set of assumptions can be used to derive more complex models, can be reduced to simpler models, or can be used to construct coupled models

Figure 4.2 illustrates the general idea of the comma categories in the context of modelling. For a selected set of assumptions (corresponding to a model), three models classes can be generated:

- the class of models derivable from this set of assumptions;
- the class of models to which the given set of assumptions can be reduced;
- the class of coupled models that contain the model corresponding to the given set of assumptions as a model being coupled.

Figure 4.2 can be seen also as an adaptation of the graphical representation of the concept of the lattice of theories discussed in Sect. 2.3. In this representation, for a selected set of assumptions we have: more complex models are located above, simpler models are located below, and coupled models are obtained by mapping the given set of assumption into a new structure. This perspective enhances understanding of a general structure of a specific modelling field and supports a better representation of relations between different models.

Finally, let us provide a simple alternative for the characterisation of universal arrows:

Definition 4.9 (Model Formulations Can Be Ordered) The ordering of sets of modelling assumptions based on model complexity is respected at the level of model formulations, even if different derivation methods have been used. Consequently, model formulations can theoretically be ordered as well; however, establishing such an ordering requires additional considerations.

It is important to emphasise once more that the definitions presented in this section capture the essence of the ideas developed within the category theory-based modelling framework presented in Chap. 3. Naturally, the rigorous constructions based on category theory are more comprehensive and address specific cases with the level of details, which is necessary for the formal description of mathematical models. The simplified definitions in this section are intended to support the practical application of the category theory-based modelling framework: we can start with these straightforward definitions and refer to the formal constructions when a deeper understanding is required. Thus, the simplified definitions serve as an interlayer between the theory and practice as indicated in Fig. 4.1.

4.2 Further Model Properties and Model Comparison

When constructing the category theory-based modelling framework in Chap. 3, we have selected model complexity, as introduced in Definition 3.2, as a *universal model property*. However, in engineering practice, other model properties, such as model uncertainty and model sensitivity, are also often used. Therefore, in this section, we explore these additional model properties and evaluate them with respect to their possible formalisation within the category theory-based modelling framework. After that, we propose categorical constructions for model comparison and discuss their practical implications.

It is important to remark, that the initial idea of using tools of abstract mathematics, and especially of category theory, for an abstract description of models and relations between them has been inspired by research conducted within the framework of the DFG GRK 1462 [7]. Consequently, when discussing model properties, we build upon the concepts developed in GRK 1462. Specifically, the following four model properties have been proposed, see [2, 3]:

- model complexity;
- model robustness;
- model sensitivity;
- model uncertainty.

While model complexity has been already addressed within the category theory-based modelling framework, the remaining three model properties—model robustness, model sensitivity, and model uncertainty—have not yet been discussed. Let us now explore these model properties and evaluate their potential for formalisation within the framework.

4.2.1 Model Robustness, Model Sensitivity, and Model Uncertainty

According to [15], model robustness focuses on analysing the response of a model with respect to variations in all possible influencing parameters, such as for example material parameters, initial and boundary conditions, and geometry. In this regard, model robustness can be seen as a generalisation of the concept of stability for initial-boundary value problems. Based on various discussions within the DFG GRK 1462, the following definition for the general concept of model robustness has been proposed:

Definition 4.10 Let X, Y, Z be Banach spaces, and $A\colon X \to Y$ and $R\colon X \to Z$ are operators, then the problem

$$\begin{cases} Au = f \text{ in } \Omega, \\ Ru = g \text{ in } \partial\Omega, \end{cases}$$

is called *robust*, if the problem is stable and if for all $\varepsilon > 0$ and for all f and g a $\delta > 0$ exists such that

$$\|A(x) - A(y)\|_Y < \varepsilon, \quad \forall x, y\colon \|x - y\|_X < \delta.$$

As we see from this definition, model robustness is directly related to model formulations rather than to the underlying modelling assumptions. A model must first be created, and only after that we could ask which mathematical structures are best suited to study the model. Therefore, it is difficult to address this model property on the abstract level within the category theory-based modelling framework. It is noteworthy that dynamical systems have been studied using category theory tools in [4], and perhaps these results can be extended to address the model robustness. However, this formalisation will still focus on model formulations rather than modelling assumptions, as required by the category theory-based modelling framework.

It is also necessary to mention another perspective on model robustness, which addresses it by a probabilistic approach, see again [2, 3]. In this approach, a model is considered robust, if its response, evaluated through stochastic simulations or sampling, has a low variance, assumed that the input variances are also small. This idea is generally aligned with the notion of model robustness introduced in Definition 4.10, but allows the model parameters be distributed probabilistically. Consequently, the probabilistic approach to model robustness also addresses the model formulation rather than modelling assumptions.

Model sensitivity is evaluated using sensitivity analysis, see [22] for details, which studies how variations in model parameters influence the model output. This analysis is based on the calculation of sensitivity indices. As discussed in Chap. 1, model sensitivity is related to the practical application of models rather than their abstract representation. Furthermore, as it has been already mentioned, it is easy to construct situations, for example by changing boundary conditions, when different input parameters will be important for the output of the same model. Consequently, the model sensitivity does not represent a universal model property.

Similar remarks can be made about model uncertainty, which characterises (stochastic) deviation of the model response from the ideal case. Analogous to model sensitivity, model formulations play a crucial role in quantifying model uncertainty. Especially parameters, such as material constants, appearing in the model formulations are particularly important for evaluating model uncertainty. Therefore, model uncertainty cannot be really addressed at the level of modelling assumption within the category theory-based modelling framework.

It is worth to emphasise, that stochastic models themselves are of course addressed by the category theory-based modelling framework, because stochastic models are nonetheless based on modelling assumptions, such as for example stochastic force is acting. Moreover, even the stochastic behaviour of material parameters can also be used as a modelling assumption. But the fact that it is necessary to perform computations to quantify the model uncertainty cannot be adequately captured by the abstract constructions presented in Chap. 3.

The discussion on model properties provided above can be summarised using the diagram shown in Fig. 4.3. Following the two-phase modelling methodology proposed in Chap. 1, we distinguish between two general classes of model properties:

- *universal model properties*, which characterise models abstractly during the conceptual modelling phase;
- *applied model properties*, which characterise models in specific practical applications and appear during the instance modelling phase.

From this subdivision into two classes, we can conclude that model robustness, model sensitivity, and model uncertainty belong to the instance modelling phase, as they directly linked to the practical use of models. In contrast, model complexity belongs to the conceptual modelling phase, as it provides an abstract characterisation of models based on their underlying modelling assumptions.

4.2.2 Model Completeness and Related Definitions

As discussed previously, model complexity, as introduced in Definition 3.2, can be seen as an abstract analogue of model fidelity. In this subsection, we make this analogy more precise by introducing another universal model property for the conceptual modelling, referred to as *model completeness*. The reason for this specification arises from the fact that model complexity is a relative measure: at least two objects are required to determine whether they are complexity-related and, if so, which one is more complex. In this context, if only a single object (set of assumptions) is provided, then it becomes impossible to evaluate it in

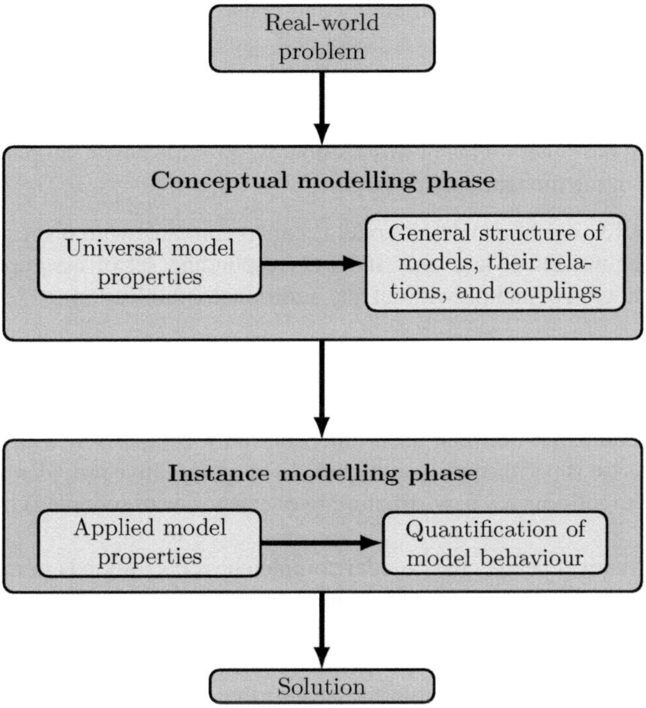

Fig. 4.3 Subdivision of model properties into two classes: universal model properties appearing during the conceptual modelling phase and applied model properties appearing during the instance modelling phase

terms of complexity. To clarify this situation, the concept of model completeness is introduced in this subsection.

We begin with the following definition introducing the notion of *model completeness* within the category theory-based modelling framework:

Definition 4.11 (Model Completeness) Let **Model** be a category of (coupled) mathematical models, and let $\mathbf{Set}_{\mathrm{A}}$ be its object together with a factorisation mapping S and model formulation A. Further, let $\mathbf{Set}'_{\mathrm{A}}$ be a set of assumptions of a certain physical phenomenon, which is not necessary an object of **Model**. Then the object $\mathbf{Set}_{\mathrm{A}}$ is called *complete* with respect to this physical phenomenon, if and only if $\mathbf{Set}_{\mathrm{A}} \subset \mathbf{Set}'_{\mathrm{A}}$, and the corresponding model formulation will also be called complete with respect to this physical phenomenon. Complete objects and models will be denoted as $\mathbf{Set}_{\mathrm{A}}\,|_{\mathbf{Set}'_{\mathrm{A}}}$ and $\mathrm{A}\,|_{\mathbf{Set}'_{\mathrm{A}}}$, respectively.

Let us now make the following remark:

Remark 4.1 Definition 4.11 is a modified version of the original definition presented in [12]. It is also important to note, that model completeness is motivated by the model theory [21], where model-completeness has been defined as follows:

Definition 4.12 Let K be a non-empty and consistent set of sentences. Whenever we mention that a structure M is a model of K we shall suppose that all relations of M are contained in K. K will be called *model-complete* if for every model M of K with diagram D, the set $K \cup D$ is complete.

Following the general idea of this section, we provide now a simple engineering formulation of the model completeness:

Definition 4.13 A mathematical model is called complete with respect to a certain physical phenomenon if and only if its corresponding set of assumptions allows describing that phenomenon without any additional modification.

As discussed above, Definition 4.11 can be viewed as a refinement of the complexity relation introduced in Definition 3.2. If $\mathbf{Set}'_{\mathrm{A}}$ is an object of the category **Model**, then completeness evidently is a special case of complexity, where not all objects are compared between each other, but only comparison with one specific object of interest is performed, namely $\mathbf{Set}'_{\mathrm{A}}$. This perspective provides an additional opportunity to introduce a new structure over categories of (coupled) mathematical models.

From a practical perspective, model completeness corresponds to model fidelity at an abstract level. In particular, we can subdivide a given category of (coupled) mathematical models into subcategories based on the physical phenomena these models can describe. Note that, in general, these subcategories may be discrete, see [1] for details. Model completeness describes the expressiveness of a model with respect to a given physical phenomenon. In other words, if a specific physical phenomenon can be described by a selected model—this is exactly the general meaning of model fidelity.

The subdivision into subcategories is particularly important for the comparative analysis of models because the objects of a category can be "filtered" with respect to a specific physical phenomenon of interest before performing a quantitative comparison. In fact, this discussion leads us to the topic of *model reduction* at the conceptual level, which, in our case, implies exactly the definition of model completeness and the respective filtering of objects in a category. Let us now formalise the concept of model reduction as follows:

Definition 4.14 (Reducible Models) Let $\mathbf{Set}_{\mathrm{A}}$ be an object of a category of (coupled) mathematical models **Model**, and let $\mathbf{Set}_{\mathrm{A}}$ be complete with respect to a certain physical phenomenon with the set of assumptions $\mathbf{Set}'_{\mathrm{A}}$. Then $\mathbf{Set}_{\mathrm{A}}$ is called *reducible*, if there is a simpler object than $\mathbf{Set}_{\mathrm{A}}$, which is still complete with respect to $\mathbf{Set}'_{\mathrm{A}}$, otherwise an object is called *irreducible*. The reduction of an object and of a model formulation are denoted then as

$$\mathbf{Set}_{\mathrm{A}} \xrightarrow[\mathbf{Set}'_{\mathrm{A}}]{} \mathbf{Set}_{\mathrm{B}}, \qquad \mathrm{A} \xrightarrow[\mathbf{Set}'_{\mathrm{A}}]{} \mathrm{B}.$$

respectively.

By reversing this definition, we can introduce the dual notion—*extendability* of a mathematical model:

Definition 4.15 (Extendable Models) Let $\mathbf{Set}_\mathrm{A}$ be an object of a category of (coupled) mathematical models **Model**, and let $\mathbf{Set}_\mathrm{A}$ be complete with respect to a certain physical phenomenon with the set of assumptions $\mathbf{Set}'_\mathrm{A}$. Then $\mathbf{Set}_\mathrm{A}$ is called *extendable*, if there is a more complex object than $\mathbf{Set}_\mathrm{A}$, which is complete with respect to $\mathbf{Set}'_\mathrm{A}$, otherwise an object is called *not-extendable*.

Finally, with the help of reducibility and extendability, we can introduce the concepts of *minimal* and *maximal* mathematical models:

Definition 4.16 (Minimal and Maximal Models) A complete object $\mathbf{Set}_\mathrm{A}$ is called *minimal* if it irreducible; similarly, a complete object $\mathbf{Set}_\mathrm{A}$ is called *maximal* if it not-extendable. Minimal and maximal model formulations are then defined analogously.

When working with models in practice, it is evidently desirable to construct minimal models, as they represent a specific physical phenomenon of interest without redundant modelling assumptions. In other words, minimal models are the simplest models capable of describing the specific physical phenomenon of interest. In that regard, identifying the minimal model is one of the tasks of the conceptual modelling phase. The category theory-based modelling framework facilitates this identification due a clear structure of categories of (coupled) mathematical models. Consequently, using minimal models also has potential to reduce computational costs during the instance modelling phase, because more complex models often require more sophisticated algorithms to work with.

Let us also make the following remark connecting the model selection process and the above discussion:

Remark 4.2 It is important to emphasise that the *model selection* is also addressed by the category theory-based modelling framework. The model selection process is reflected in the choice of a specific physical phenomenon of interest, $\mathbf{Set}'_\mathrm{A}$. Thus, the framework enables the addressing of model selection and model reduction at the conceptual level, even though these topics are typically considered only in the context of concrete computations. This highlights another advantage of working with category theory and abstract structures.

Let us conclude this subsection by illustrating the ideas presented above with an example from the category **Heat**. The sets of assumptions corresponding to the objects of this category are presented in Table 3.3. Now, let us assume that we are interested in describing the following physical phenomenon:

$$\mathbf{Set}'_\mathrm{F} = \{\text{Heat propagates in a body according to Fourier's law}\}\,.$$

In this case, all objects of the category **Heat** are complete with respect to $\mathbf{Set}'_\mathrm{F}$. Further, we can list the following facts:

- Object $\mathbf{Set}_{\mathrm{UH}}$ is the minimal object with respect to $\mathbf{Set}'_{\mathrm{F}}$;
- Object $\mathbf{Set}_{\mathrm{NURH}}$ is the maximal object with respect to $\mathbf{Set}'_{\mathrm{F}}$;
- Objects $\mathbf{Set}_{\mathrm{UH}}$, $\mathbf{Set}_{\mathrm{NUH}}$, $\mathbf{Set}_{\mathrm{RH}}$ are extendable;
- Objects $\mathbf{Set}_{\mathrm{NURH}}$, $\mathbf{Set}_{\mathrm{NUH}}$, $\mathbf{Set}_{\mathrm{RH}}$ are reducible.

This example illustrates the trivial case, when we are interested in a physical phenomenon, which belongs to every object of a category. Hence, model reduction can be easily performed by selecting the initial object of this category $\mathbf{Set}_{\mathrm{UH}}$.

Let us now consider a more interesting case by addressing the following set of assumptions:

$$\mathbf{Set}'_{\mathrm{RN}} = \left\{ \begin{array}{l} \text{Body is made of a non-homogeneous material} \\ \text{There is a radiative heat exchange with the surroundings} \end{array} \right\}.$$

In this case, only object $\mathbf{Set}_{\mathrm{NURH}}$ is complete with respect to $\mathbf{Set}'_{\mathrm{RN}}$, implying that this object simultaneously minimal and maximal object of this category. Thus, no model reduction is possible in this case, and we have to work with the terminal object of the category.

4.2.3 Comparison of Models at the Abstract Level

Let us now address the question of model comparability from the practical point of view. On the level of categories, models can be ordered, and therefore compared, by using the complexity relation. Given that various constructions for the complexity of coupled models have been introduced in Sect. 3.3.2, it is worth beginning the discussion on the practical comparability of models by clearly stating what can be compared at the categorical level:

- Coupled models can be compared with respect to the complexity of their coupling conditions.
- Further, coupled models can be compared on the left and on the right, meaning that mathematical models, associated with the objects of categories being coupled, are compared.

In practical applications, comparison of models should involve not only ordering them with respect to a selected criterion but also the quantitative evaluation of their responses. For now on, we use the term *System Response Quantity* (SRQ) introduced in [19], to denote the output of the modelling process. In this context, the practical comparison of models involves evaluating the influence of specific modelling assumptions on a selected SRQ, such as for example the displacement of a system or the maximum temperature in a body.

As discussed above, (coupled) mathematical models can be compared by using the complexity relation, which allows to identify specific modelling assumptions influencing the model response. In practice, however, it is often required to compare models that are not directly complexity-related. For example, if we consider the category **Heat** and aim to compare a heat propagation model for a non-homogeneous material with and a heat propagation model for radiative exchange with the surroundings, then this comparison cannot be performed using the complexity relation, because the corresponding objects, $\mathbf{Set}_{\mathrm{NUH}}$ and $\mathbf{Set}_{\mathrm{RH}}$, are not complexity-related. To overcome this problem, the notion of *relative comparison of models* has been introduced in [12]. Let us now propose a refined version of this Definition:

Definition 4.17 (Relative Comparison of Models) Let $\mathbf{Set}_{\mathrm{A}}$, $\mathbf{Set}_{\mathrm{B}}$, and $\mathbf{Set}_{\mathrm{C}}$ be objects of the category of (coupled) mathematical models **Model**, such that one of the following diagrams holds

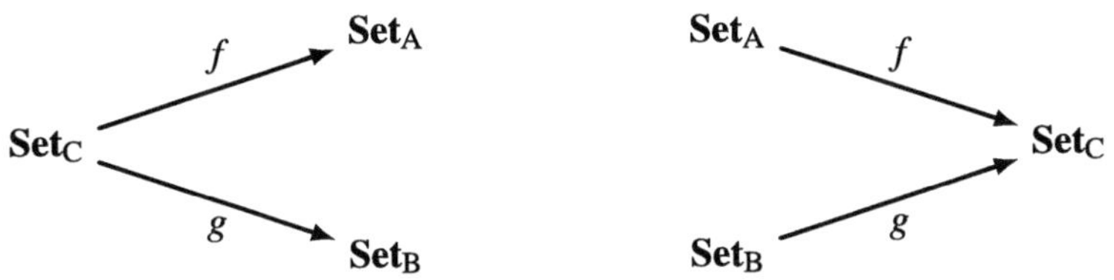

Then objects $\mathbf{Set}_{\mathrm{A}}$ and $\mathbf{Set}_{\mathrm{B}}$ are called *relatively comparable* with respect to object $\mathbf{Set}_{\mathrm{C}}$, if and only if one of the following conditions holds

$$(\mathbf{Set}_{\mathrm{A}} \cup \mathbf{Set}_{\mathrm{B}}) \subseteq \mathbf{Set}_{\mathrm{C}} \quad \text{or} \quad \mathbf{Set}_{\mathrm{C}} \subseteq (\mathbf{Set}_{\mathrm{A}} \cup \mathbf{Set}_{\mathrm{B}}),$$

respectively. Similarly, the corresponding (coupled) mathematical models are called *relatively comparable*.

The concept of relative comparability of models, introduced in Definition 4.17, can be interpreted as an analogy to the notion of a *common ground* when comparing two different worldviews in the context of ontological representation of knowledge, see [23] for details. In this context, a quantitative comparison of two objects, $\mathbf{Set}_{\mathrm{A}}$ and $\mathbf{Set}_{\mathrm{B}}$, which are not complexity-related, requires the introduction of a third object, $\mathbf{Set}_{\mathrm{C}}$, as a common ground connecting the two objects. Consequently, the object $\mathbf{Set}_{\mathrm{C}}$ serves as an interface and is either more complex or simpler than the other two objects.

Additionally, the relative comparison of models provide the means to compare objects in a partially ordered category that are not directly related. Nonetheless, it is important to make the following remark:

Remark 4.3 When using the relative comparison of models in practice, it is important to keep in mind that if a simpler model is used for the comparison, then it is not possible to conclude which model's outcome is of higher quality. In this case, only the effects of excluding specific assumptions from the set of assumptions of the simpler model can be studied.

Finally, let us look at more practical aspects of comparing models by studying the effect of modelling assumptions on a selected SRQ. Since the goal of such a comparison is to perform a quantitative evaluation, a *comparison metric* needs to be defined, such as for example the mean square error or L^2-error. Furthermore, any comparison requires a reference model, based on which the relative effect of excluding or adding an assumption with respect to a selected SRQ is studied. Depending on the reference model chosen, we can distinguish between *forward* and *backward* comparison, defined as follows [12]:

Definition 4.18 (Forward and Backward Comparison) Let a comparison metric **M** quantify the difference of a SRQ between the models A and B, where A is chosen to be the reference model. We say that we conduct a:

(i) *forward comparison*, if $\mathbf{Set}_\mathrm{B} \subseteq \mathbf{Set}_\mathrm{A}$, and we denote the comparison metric M as $\mathbf{M}^{\mathrm{A,B}}_{SRQ}$;
(ii) *backward comparison*, if $\mathbf{Set}_\mathrm{A} \subseteq \mathbf{Set}_\mathrm{B}$, and we denote the comparison metric M as $\overline{\mathbf{M}^{\mathrm{A,B}}_{SRQ}}$.

We conclude this subsection with the following remark:

Remark 4.4 In fact, comma categories of the form $(\mathbf{Set}_\mathrm{l} \downarrow \mathbf{Model})$ can be seen as categories providing *forward comparability of mathematical models*, while comma categories of the form $(\mathbf{Model} \downarrow \mathbf{Set}_\mathrm{l})$ can be seen as categories providing *backward comparability of mathematical models.*

4.3 Categorical Approach to Aerodynamic Analysis of Bridge Decks

The goal of this section is to illustrate how the category theory-based modelling framework can be used in real-world engineering applications. To this end, we will briefly revisit the problem of aerodynamic analysis of bridge decks presented in [12]. Because our focus here is on discussing the applications and benefits of categorical constructions in practical engineering, we will only provide a brief overview of aerodynamic models and the concept of fluid-structure interaction. For detailed information on the aerodynamic analysis of bridges, we refer to the works [10, 11, 13, 14] and references therein.

The problem of fluid-structure interaction in the context of the bridge aerodynamics consist of modelling a deformable structure (bridge deck) under the influence of wind forces, see Fig. 4.4. To account for the motion of a structure, we consider a liner structural model S, which is given by the following model formulation:

$$\mathrm{S} := \begin{cases} m_h \ddot{h} + c_h \dot{h} + k_h h = L, \\ m_\alpha \ddot{\alpha} + c_\alpha \dot{\alpha} + k_\alpha \alpha = M, \end{cases} \tag{4.1}$$

Fig. 4.4 A simplified system to model bridge deck with chord B under wind vibrations by using semi-analytic models

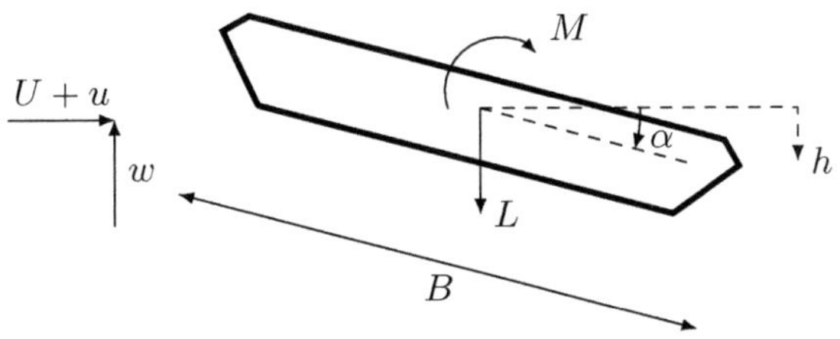

where k_h and k_α are the vertical and rotational spring, respectively; h is the vertical displacement, α alpha is the rotational displacement; m_h and m_α are the inertial mass and moment; c_h and c_k are the vertical and rotational structural damping coefficients; $L = L(t)$ is the time-dependent lift force, and $M = M(t)$ is the moment force.

Let us now list the model formulations of eleven semi-analytical aerodynamic models:

1. Steady model:

$$\mathrm{ST} := \begin{cases} L = F_L \cos\phi - F_D \sin\phi, \; F_D = \frac{1}{2}\rho U_r^2 B C_D(\alpha_e), \\ F_L = -\frac{1}{2}\rho U_r^2 B C_L(\alpha_e), \quad M = \frac{1}{2}\rho U_r^2 B^2 C_M(\alpha_e), \\ \alpha_e = \alpha_s + \phi, \qquad \phi = \arctan\left(\frac{w}{U+u}\right), \\ U_r = \sqrt{(U+u)^2 + w^2}, \end{cases}$$

where U_r is the resultant wind velocity considering only the wind fluctuations, α_e is the effective angle of attack, ϕ is the dynamic angle of attack, and α_s is the wind angle of attack at equilibrium at equilibrium position. The drag, lift and moment static wind coefficients are denoted as C_D, C_L, and C_M, respectively.

2. Linear steady model:

$$\mathrm{LST} := \begin{cases} L = -\frac{1}{2}\rho U^2 B\left[C_L + 2C_L\frac{u}{U} + (C_L' + C_D)\frac{w}{U}\right], \\ M = \frac{1}{2}\rho U^2 B^2\left[C_M + 2C_M\frac{u}{U} + C_M'\frac{w}{U}\right], \end{cases}$$

where C_D=$C_D(\alpha_s)$, C_L=$C_L(\alpha_s)$, and C_M=$C_M(\alpha_s)$ are the static wind coefficients for α_s.

3. Quasi-steady model:

$$
\mathrm{QS} := \begin{cases} L = F_L \cos\phi - F_D \sin\phi, \qquad F_D = \frac{1}{2}\rho U_r^2 B C_D(\alpha_e), \\ F_L = -\frac{1}{2}\rho U_r^2 B C_L(\alpha_e), \qquad M = \frac{1}{2}\rho U_r^2 B^2 C_M(\alpha_e), \\ \alpha_e = \alpha_s + \alpha + \phi, \qquad \phi = \arctan\left(\frac{w + \dot{h} + m_1 B\dot{\alpha}}{U + u}\right), \\ U_r = \sqrt{(U+u)^2 + (w + \dot{h} + m_1 B\dot{\alpha})^2}, \end{cases} \tag{4.2}
$$

where the m_1 coefficient specifies the position of the aerodynamic centre.

4. Linear quasi-steady model:

$$
\mathrm{LQS} := \begin{cases} L = -\frac{1}{2}\rho U^2 B \Big[C_L + 2C_L \frac{u}{U} + (C_L' + C_D)\frac{w}{U} \\ \qquad + (C_L' + C_D)\frac{\dot{h} + m_1 B\dot{\alpha}}{U} + C_L'\alpha \Big], \\ M = \frac{1}{2}\rho U^2 B^2 \Big[C_M + 2C_M \frac{u}{U} + C_M' \frac{w}{U} \\ \qquad + C_M' \frac{\dot{h} + m_1 B\dot{\alpha}}{U} + C_M'\alpha \Big]. \end{cases} \tag{4.3}
$$

5. Linear unsteady model:

$$
\mathrm{LU} := \begin{cases} L = -\frac{1}{2}\rho U^2 B \Big\{ C_L + 2C_L \int_0^t \Phi_{Lu}(t-\tau)\frac{\dot{u}(\tau)}{U}\mathrm{d}\tau \\ \quad + \left(C_L' + C_D\right) \int_0^t \Big[\Phi_{Lw}(t-\tau)\frac{\dot{w}(\tau)}{U} + \Phi_{Lh}(t-\tau)\frac{\ddot{h}(\tau)}{U} \\ \quad + \Phi_{L\dot{\alpha}}(t-\tau)\frac{B\ddot{\alpha}(\tau)}{U} + \Phi_{L\alpha}(t-\tau)\dot{\alpha}(\tau) \Big] \mathrm{d}\tau \Big\}, \\ M = \frac{1}{2}\rho U^2 B^2 \Big\{ C_M + 2C_M \int_0^t \Phi_{Mu}(t-\tau)\frac{\dot{u}(\tau)}{U}\mathrm{d}\tau \\ \quad + C_M' \int_0^t \Big[\Phi_{Mw}(t-\tau)\frac{\dot{w}(\tau)}{U} + \Phi_{Mh}(t-\tau)\frac{\ddot{h}(\tau)}{U} \\ \quad + \Phi_{M\dot{\alpha}}(t-\tau)\frac{B\ddot{\alpha}(\tau)}{U} + \Phi_{M\alpha}(t-\tau)\dot{\alpha}(\tau) \Big] \mathrm{d}\tau \Big\}, \end{cases} \tag{4.4}
$$

where Φ_L, and Φ_M are the lift and moment indicial functions, respectively.

6. Modified quasi-steady model:

$$\mathrm{MQS} := \begin{cases} L = -\frac{1}{2}\rho U^2 B \left[C_L + 2C_L \frac{u}{U} + (C_L^{'} + C_D)\frac{w}{U} - h_1 \frac{\dot{h}}{U} \right. \\ \qquad \left. -h_2 \frac{B\dot{\alpha}}{U} - h_3 \alpha - h_4 \frac{h}{B} \right], \\ M = \frac{1}{2}\rho U^2 B^2 \left[C_M + 2C_M \frac{u}{U} + C_M^{'} \frac{w}{U} + a_1 \frac{\dot{h}}{U} + a_2 \frac{B\dot{\alpha}}{U} \right. \\ \qquad \left. +a_3 \alpha + a_4 \frac{h}{B} \right], \end{cases} \tag{4.5}$$

where $h_j = K_c H_j^*(K_c)$, $a_j = K_c A_j^*(K_c)$ for $j = \{1, 2\}$ and $h_j = K_c^2 H_j^*(K_c)$, $a_j = K_c^2 A_j^*(K_c)$ for $j = \{3, 4\}$ are the frequency-independent coefficients taking into account the average fluid memory at a specific reduced frequency of oscillation K_c.

7. Mode-by-mode model:

MBM

$$:= \begin{cases} L = -\frac{1}{2}\rho U^2 B \left\{ C_L + \int_0^t 2C_L \Phi_{Lu}(t-\tau)\frac{\dot{u}(\tau)}{U}\mathrm{d}\tau \right. \\ \qquad \left. + \left(C_L^{'} + C_D\right) \int_0^t \left[\Phi_{Lw}(t-\tau)\frac{\dot{w}(\tau)}{U} + \Phi_{Lh}(t-\tau)\frac{\ddot{h}(\tau)}{U} \right] \mathrm{d}\tau \right\}, \\ M = \frac{1}{2}\rho U^2 B^2 \left\{ C_M + \int_0^t 2C_M \Phi_{Mu}(t-\tau)\frac{\dot{u}(\tau)}{U}\mathrm{d}\tau \right. \\ \qquad + \int_0^t C_M^{'} \left[\Phi_{Mw}(t-\tau)\frac{\dot{w}(\tau)}{U} + \Phi_{M\dot{\alpha}}(t-\tau)\frac{B\ddot{\alpha}(\tau)}{U} \right. \\ \qquad \left. \left. +\Phi_{M\alpha}(t-\tau)\dot{\alpha}(\tau) \right] \mathrm{d}\tau \right\}. \end{cases}$$

8. Corrected quasi-steady model:

$$\mathrm{CQS} := \begin{cases} L = F_L \cos\phi - F_D \sin\phi, \quad F_D = \frac{1}{2}\rho U_r^2 B C_D^*(\alpha_e), \\ F_L = -\frac{1}{2}\rho U_r^2 B C_L^*(\alpha_e), \quad M = \frac{1}{2}\rho U_r^2 B^2 C_M^*(\alpha_e), \\ \alpha_e = \alpha_s + \alpha + \phi, \quad \phi = \arctan\left(\frac{w + \dot{h} + m_1 B\dot{\alpha}}{U + u} \right), \\ U_r = \sqrt{(U+u)^2 + (w + \dot{h} + m_1 B\dot{\alpha})^2}, \\ C_j^*(\alpha_e) = C_j(\alpha_s) + \int_{\alpha_s}^{\alpha_e} K_j^*(\alpha, \omega_c) C_j^{'}(\alpha)\mathrm{d}\alpha, \end{cases} \tag{4.6}$$

for $j = \{D, L, M\}$.

9. Hybrid nonlinear model:

$$\mathrm{HNL} := \begin{cases} L = L^{QS}(\alpha_e^l) + L^{LU}(\alpha_e^h)|_{\alpha_e^l}, \quad M = M^{QS}(\alpha_e^l) + M^{LU}(\alpha_e^h)|_{\alpha_e^l}, \\ \alpha_e = \alpha_e^l + \alpha_e^h, \\ \alpha_e^l = \alpha_s + \alpha^l + \arctan\left(\dfrac{w^l + \dot{h}^l + m_1 B\dot{\alpha}^l + n_1\dot{w}^l}{U + u^l}\right), \end{cases} \tag{4.7}$$

where $|_{\alpha_e^l}$ denotes linearisation at α_e^l, which includes the small angle hypothesis and disregards higher order terms.

10. Modified nonlinear model:

$$\mathrm{MNL} := \begin{cases} L = -\dfrac{1}{2}\rho U_r^2 B\left[C_L^{hys}(\alpha_e, \dot{\alpha}_e, \omega_c)\cos\phi + C_D^{hys}(\alpha_e, \dot{\alpha}_e, \omega_c)\sin\phi\right], \\ M = \dfrac{1}{2}\rho U_r^2 B^2 C_M^{hys}(\alpha_e, \dot{\alpha}_e, \omega_c), \quad \alpha_e = \alpha_s + \alpha + \phi, \\ \phi = \arctan\left(\dfrac{w + \dot{h} + m_1 B\dot{\alpha}}{U + u}\right), \\ U_r = \sqrt{(U + u)^2 + (w + \dot{h} + m_1 B\dot{\alpha})^2}, \\ C_j^{hys} = C_j(\alpha_s) + \sum_{k,l}^{n} \eta_{jkl}\alpha_e^k(\omega_c)\dot{\alpha}_e{}^l(\omega_c), \end{cases} \tag{4.8}$$

for $j = \{D, L, M\}$ and where $C^{hys} = C^{hys}(\alpha_e, \dot{\alpha}_e, \omega_c)$ is the aerodynamic hysteresis and η is the approximation coefficient.

11. Nonlinear unsteady model:

$$\mathrm{NLU} := \begin{cases} L = -\dfrac{1}{2}\rho U^2 B\Big\{\Phi_L^0 + \sum_k \int_0^t \Phi_{Lk}^{\mathrm{I}}(t-\tau)\dot{k}(\tau)\mathrm{d}\tau \\ \quad + \sum_{k,l}\int_0^t\int_0^t \Phi_{Lkl}^{\mathrm{II}}(t-\tau_1, t-\tau_2)\dot{k}(\tau_1)\dot{l}(\tau_2)\mathrm{d}\tau_1\mathrm{d}\tau_2 + \ldots\Big\}, \\ M = \dfrac{1}{2}\rho U^2 B^2\Big\{\Phi_M^0 + \sum_k \int_0^t \Phi_{Mk}^{\mathrm{I}}(t-\tau)\dot{k}(\tau)\mathrm{d}\tau \\ \quad + \sum_{k,l}\int_0^t\int_0^t \Phi_{Mkl}^{\mathrm{II}}(t-\tau_1, t-\tau_2)\dot{k}(\tau_1)\dot{l}(\tau_2)\mathrm{d}\tau_1\mathrm{d}\tau_2 + \ldots\Big\}, \end{cases} \tag{4.9}$$

for $k, l = \{u, w, \dot{h}, \alpha, \dot{\alpha}\}$, and where Φ^0, Φ^{I}, and Φ^{II} are the zeroth, first and second order indicial functions.

Let us now denote by **AeroModel** the category of aerodynamic models, whose formulations have been introduced above. Note that formally these model formulations are objects of the category $\mathbf{AeroModel}^F$, while the objects of the category **AeroModel** are sets of assumptions of these models, i.e. $\mathbf{Set}_{\mathrm{ST}}$, $\mathbf{Set}_{\mathrm{LST}}$ etc. Category **AeroModel** has the following structure, see [12, 17]:

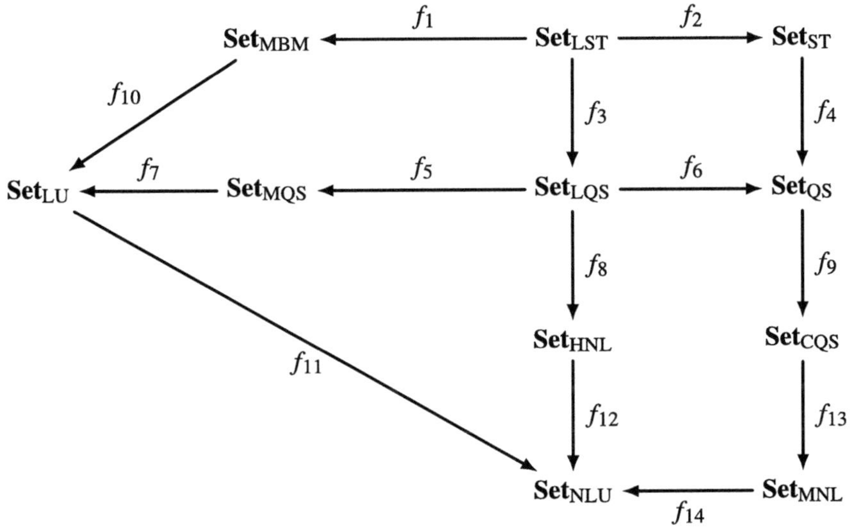

It is necessary to make the following remark:

Remark 4.5 Note that in work [12], the above diagram has been created for model formulations, rather than for the objects of **AeroModel**, because ordering of models has been based on formalised sets of assumptions rather sentences formulated in natural language. This approach is, of course, also a valid way to work with the sets of assumptions and utilise the category theory-based modelling framework. In fact, the framework provides a certain level of flexibility: the modeller can decide which approach best suits their current needs—describe sets of assumptions in a natural language, or consider their formalised versions. At the same time, all possible constructions remain mathematically sound, and there is always a clear way how to switch between the two options.

Let us now summarise known facts about the category **AeroModel**:

- **AeroModel** is a partially ordered category;
- The object $\mathbf{Set}_{\mathrm{LST}}$ is the initial object of this category;
- The object $\mathbf{Set}_{\mathrm{NLU}}$ is the terminal object of this category;
- According to Theorem 3.1, five totally ordered subcategories exists, which are shown in Fig. 4.5.

1. $\mathbf{Set}_{\mathrm{LST}} \xrightarrow{f_2} \mathbf{Set}_{\mathrm{ST}} \xrightarrow{f_4} \mathbf{Set}_{\mathrm{QS}} \xrightarrow{f_9} \mathbf{Set}_{\mathrm{CQS}} \xrightarrow{f_{13}} \mathbf{Set}_{\mathrm{MNL}} \xrightarrow{f_{14}} \mathbf{Set}_{\mathrm{NLU}}$
2. $\mathbf{Set}_{\mathrm{LST}} \xrightarrow{f_3} \mathbf{Set}_{\mathrm{LQS}} \xrightarrow{f_6} \mathbf{Set}_{\mathrm{QS}} \xrightarrow{f_9} \mathbf{Set}_{\mathrm{CQS}} \xrightarrow{f_{13}} \mathbf{Set}_{\mathrm{MNL}} \xrightarrow{f_{14}} \mathbf{Set}_{\mathrm{NLU}}$
3. $\mathbf{Set}_{\mathrm{LST}} \xrightarrow{f_3} \mathbf{Set}_{\mathrm{LQS}} \xrightarrow{f_8} \mathbf{Set}_{\mathrm{HNL}} \xrightarrow{f_{12}} \mathbf{Set}_{\mathrm{NLU}}$
4. $\mathbf{Set}_{\mathrm{LST}} \xrightarrow{f_3} \mathbf{Set}_{\mathrm{LQS}} \xrightarrow{f_5} \mathbf{Set}_{\mathrm{MQS}} \xrightarrow{f_7} \mathbf{Set}_{\mathrm{LU}} \xrightarrow{f_{11}} \mathbf{Set}_{\mathrm{NLU}}$
5. $\mathbf{Set}_{\mathrm{LST}} \xrightarrow{f_1} \mathbf{Set}_{\mathrm{MBM}} \xrightarrow{f_{10}} \mathbf{Set}_{\mathrm{LU}} \xrightarrow{f_{11}} \mathbf{Set}_{\mathrm{NLU}}$

Fig. 4.5 Totally ordered subcategories of the category **AeroModel**

If we are now interested in aerodynamic models, which are able to describe the classical flutter phenomenon, see again [14] for details, then the category **FlutterModel** containing objects, which are complete with respect to the classical flutter, can be created. Evidently, this category is a subcategory of **AeroModel**, and it is formally obtained by help of the inclusion functor:

$$\mathbf{FlutterModel} \xrightarrow{I} \mathbf{AeroModel}.$$

The category **FlutterModel** contains then the following objects

$$\mathbf{Set}_{\mathrm{LQS}}, \mathbf{Set}_{\mathrm{MQS}}, \mathbf{Set}_{\mathrm{QS}}, \mathbf{Set}_{\mathrm{LU}}, \mathbf{Set}_{\mathrm{HNL}}, \mathbf{Set}_{\mathrm{CQS}}, \mathbf{Set}_{\mathrm{MNL}}, \mathbf{Set}_{\mathrm{NLU}},$$

and has the structure represented by the diagram

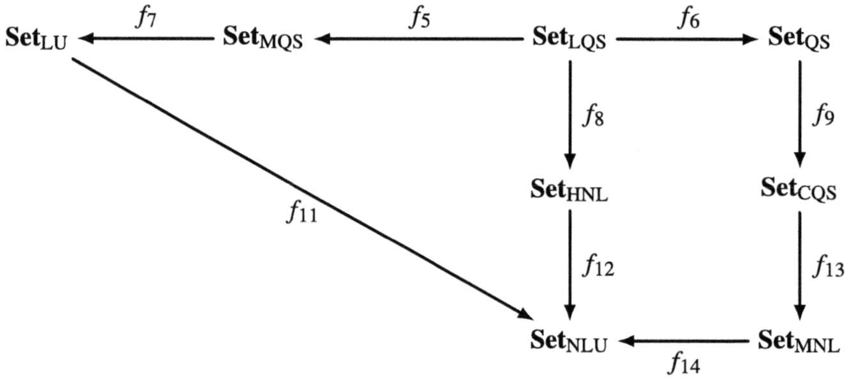

For this category, we can conclude the following:

- **FlutterModel** is a partially ordered category;
- The object $\mathbf{Set}_{\mathrm{LQS}}$ is the initial object of this category;
- The object $\mathbf{Set}_{\mathrm{NLU}}$ is the terminal object of this category;
- The object $\mathbf{Set}_{\mathrm{LQS}}$ is the minimal object of this category;
- The object $\mathbf{Set}_{\mathrm{NLU}}$ is the maximal object of this category;

- Evidently, the category **FlutterModel** again contains several totally ordered subcategories, which we will not list explicitly here.

Finally, let us illustrate how comparison metrics can be applied in practical applications by revisiting the example of the Great Belt Bridge presented in [12]. The structure of the category **AeroModel** clearly indicates which models can be compared. For illustrative purposes, let us now choose the forward comparison metric based on the root-mean-square of the vertical response. This metric is defined as follows

$$M_{h_{rms}}^{\mathrm{A,B}} := \frac{h_{rms}^{\mathrm{A}} - h_{rms}^{\mathrm{B}}}{h_{rms}^{\mathrm{A}}},$$

where the model A has higher complexity than the model B. Correspondingly, the backward comparison metric is denoted as $\overline{M_{h_{rms}}^{\mathrm{A,B}}}$. A comparison metric for the rotation is defined analogously:

$$M_{\alpha_{rms}}^{\mathrm{A,B}} := \frac{\alpha_{rms}^{\mathrm{A}} - \alpha_{rms}^{\mathrm{B}}}{\alpha_{rms}^{\mathrm{A}}}, \text{ and } \overline{M_{\alpha_{rms}}^{\mathrm{A,B}}}$$

The model comparison can now be performed for various models associated with the objects of **AeroModel**. For instance, the effect of aerodynamic nonlinearity in the models QS and LQS can be studied through concrete calculations. The objects $\mathbf{Set}_{\mathrm{QS}}$ and $\mathbf{Set}_{\mathrm{LQS}}$ are complexity-related with the object $\mathbf{Set}_{\mathrm{QS}}$ has higher complexity, as indicated by the morphism f_6 in the diagram depicting structure of the category. By analysing the sets of assumptions of the models QS and LQS, it can be observed that the principal difference lies in the linearisation. Consequently, we can study the effect of nonlinearity, based on the set of assumptions $\mathbf{Set}_{\mathrm{QS}}$, for which $M_{rms}^{\mathrm{QS,LQS}}$ is shown in Fig. 4.6 (left). The figure indicates that including aerodynamic nonlinearity, based on the QS model, reduces the root-mean-square of the response. Particularly, for the rotation, the overestimation for the LQS model increases for higher wind velocities.

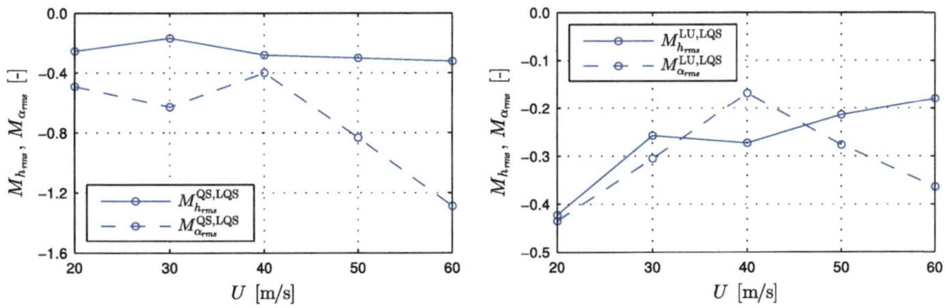

Fig. 4.6 Relative comparative metric based on the root-mean square [12]

Further, by comparing the LU and LQS models, the effect of linear fluid memory on the SRQ can be analysed. In this case, the object $\mathbf{Set}_{\mathrm{LU}}$ has higher complexity than object $\mathbf{Set}_{\mathrm{LQS}}$, and the formalised assumption of interest is

$$\mathbf{Set}_{\mathrm{LQS}} \setminus \mathbf{Set}_{\mathrm{LU}} = \{V_r \to \infty\},$$

based on $\mathbf{Set}_{\mathrm{LU}}$. Figure 4.6 (right) shows the relative comparison metric $M_{rms}^{\mathrm{LU,LQS}}$. From the figure, it can be concluded that including the linear fluid memory based on the LU model reduces the response. The difference decreases with increasing wind speeds (thus, increasing V_r) for the vertical degree of freedom, as the effect of fluid memory should be insignificant at high V_r. However, the difference for the rotation for $U > 40$ m/s increases, which is attributed to the effect of fluid memory on the aerodynamic coupling [14]. In [12], further comparison and analysis of models has been performed, which underline that abstract construction defined at the level of categories can be transferred to concrete engineering applications and numerical calculations.

Let us conclude this section with the following remark:

Remark 4.6 The practical applications presented in this section emphasises once more that the category theory-based modelling framework proposed in Chap. 3 addresses the conceptual modelling phase and can effectively support a detailed analysis of models in the instance modelling phase. It is necessary to underline that practical problems cannot be solve solely by abstract methods, but their proper treatment requires a combination of abstract approaches and concrete calculations, as shown in Fig. 4.3.

4.4 Categorical Perspective on the Development of Coupled Solution Methods

In this section, we discuss another practical application of the category theory-based modelling framework, specifically the development of solution methods for coupled problems. The overall goal here is present ideas on checking *consistency* of a numerical scheme used to solve a coupled problem, or numerical schemes coupling different solution methods.

As introduced in Definitions 3.6–3.7, coupled models always include specific coupling conditions. By using the category theory-based modelling framework, these coupling conditions can be clearly formulated during the conceptual modelling phase. Consequently, the coupling conditions must be also appropriately addressed during the instance modelling phase when solving coupled problems in practice.

Let us now try to make more formal the idea of a consistent treatment of coupling conditions in practice. The practical use of models typically results in solving an initial-boundary value problem for a specific model formulation. Hence, we can represent the use of models in practice by the following scheme:

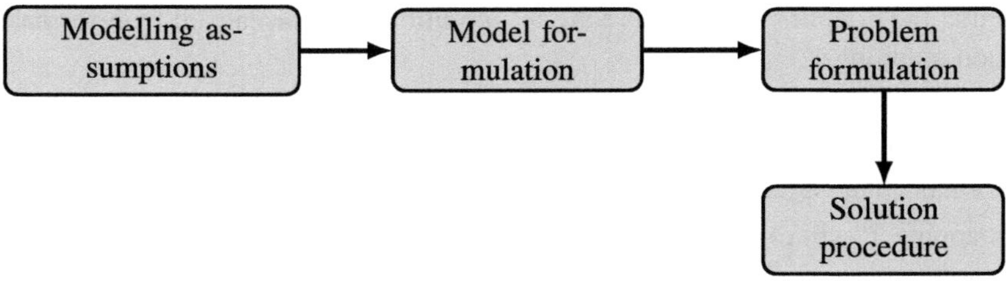

The term "solution procedure" is understood here in a broad sense, encompassing both analytical and numerical solutions methods. The reason for this broad interpretation is that we want to express the general notion of consistency in the connection between solution procedure and modelling assumptions, rather than to discuss specific algorithmic details.

Let us now discuss the arrows in the above diagram:

- The arrow between "Modelling assumptions" and "Model formulation" is, in fact, a formalisation mapping S from Definition 3.1, which formalises modelling assumptions in terms of mathematical expressions and equations.
- The arrow between "Model formulation" and "Problem formulation" is a mapping, which adds boundary and/or initial conditions to the model formulation. For the sake of this discussion, we will refer to this mapping as a *formulation mapping*. The "Problem formulation" block is necessary because the category theory-based modelling framework treats models independently of specific boundary or initial conditions, which are required when solving practical problems.
- Finally, the arrow between "Problem formulation" and "Solution procedure" can be interpreted as a mapping, which maps a given initial-boundary value problem to a solution method. Analogously to the formulation mapping, we will refer to this mapping as a *solution procedure mapping*.

A categorical formalisation of the formulation and solution procedure mappings requires appropriate categorical definitions for "Problem formulation" and "Solution procedure", meaning that categories describing these general blocks need to be defined. However, it is unclear whether such definitions could be meaningfully introduced. Therefore, we will not attempt to develop them here. Instead, we focus on discussing the mapping of coupling conditions, as this allows for more precise statements regarding consistency.

Let $\mathbf{Set}_{ij}^{\mathrm{Coup}}$ denote the set of coupling conditions of a coupled mathematical model. Moreover, to cover the cases of coupling of models from different categories and from the same category, we allow i to be equal j. Further, let $\mathbf{Sol}_{ij}^{\mathrm{Coup}}$ denote the subpart of a general solution procedure, representing the coupling conditions used in the solution procedure. Let us now analyse step-by-step how the coupling conditions $\mathbf{Set}_{ij}^{\mathrm{Coup}}$ are mapped through the general scheme presented above:

1. First, the coupling conditions $\mathbf{Set}_{ij}^{\mathrm{Coup}}$ conditions are formalised by a formalisation mapping S.
2. Second, the formalised coupling conditions $S\left(\mathbf{Set}_{ij}^{\mathrm{Coup}}\right)$ are extended by a formulation mapping T, which adds boundary and/or initial conditions to a formalised coupled mathematical model. Note, that in practical applications, the mapping T will often act as identity mapping on $S\left(\mathbf{Set}_{ij}^{\mathrm{Coup}}\right)$, because coupling conditions are typically not affected by boundary and/or initial conditions.
3. Third, the object $\mathbf{Sol}_{ij}^{\mathrm{Coup}}$ is constructed by applying a solution procedure mapping P to the outcome of step 2. Hence, we have the following representation

$$\mathbf{Sol}_{ij}^{\mathrm{Coup}} = (P \circ T \circ S)\, \mathbf{Set}_{ij}^{\mathrm{Coup}}. \tag{4.10}$$

In practical applications, we can characterise the coupling as *consistent* if Eq. (4.10) is satisfied exactly, and as inconsistent otherwise. The exact satisfaction of (4.10) is understood here in a broader sense. For instance, if we consider again the heat conduction problem in a domain composed of two different materials, then the coupling conditions are:

1. the temperature field must be continuous, and
2. the heat flux must be continuous,

at the coupling interface. Consequently, if, during a solution procedure, the continuity of the temperature field is substituted by an L^2-approximation at the interface without ensuring sufficient regularity conditions for the solution to also belong to the C-space, then the coupling is realised inconsistently with respect to the model. In this case, we say that (4.10) is not satisfied exactly.

It is important to make the following remark:

Remark 4.7 The ideas presented in this section are related to the theoretical aspects of modelling. In this regard, it may happen that it is not possible to realise a consistent coupling in practice due to the limitations of current computational algorithms, knowledge, or other constraints. Such situations are typical because the development of modern technologies often requires algorithms and ideas that are not yet fully understood or developed from a theoretical perspective. Thus, the of this section is not to label all such approaches as incorrect, but rather to clearly highlight the limitations when working with models in practice. The final decision is nonetheless made by the modeller.

The above discussion fully covers the case of coupling models from different categories and, consequently, the development of coupled solution procedure and checking their consistency with respect to the coupling. Additionally, coupled solution procedures are often used when solving boundary value problems with singularities. The presence of singularity requires specific treatment in constructing

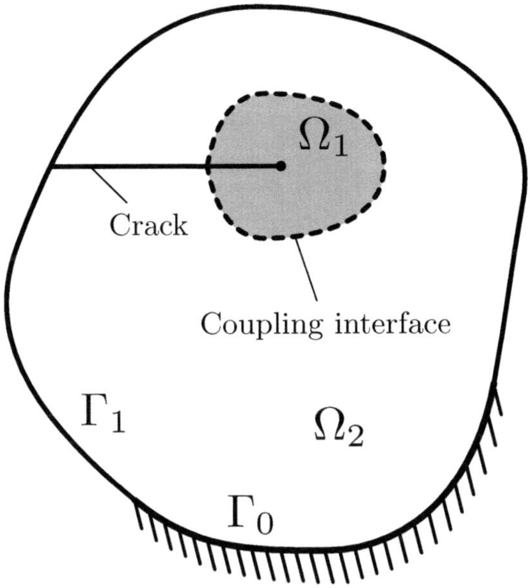

Fig. 4.7 An example of using a coupled solution method in fracture mechanics: Ω_1 is the domain near the singularity, Ω_2 is the domain far from the singularity, Γ_0 is the part of the boundary with Dirichlet boundary conditions, and Γ_1 is the part of the boundary with Neumann boundary conditions. Different solutions methods used in Ω_1 and Ω_2 are then coupled together

the solution to accurately represent the singular behaviour. Such singular problems are typically not consider as coupled problems, because the coupling appears only at the level of the solution procedures. However, as we will show later in this section, changing of perspective can provided a clearer understanding of the coupling conditions that must be satisfied when developing coupled solution procedures.

We will now discuss the classical boundary value problem from a two-dimensional linear elastic fracture mechanics of calculating a stress field within a domain Ω containing a crack, see [18]. The singularity appears because of the crack tip, see [8] for details. Using this example, we will illustrate how the category theory-based modelling framework can support the development of coupled solution procedures in practice.

Let us consider the domain Ω shown in Fig. 4.7 containing a crack. This domain is further divided into two subdomains: Ω_1, which surrounds the crack tip, and Ω_2, representing the region far from the crack tip singularity. A typical approach to the solution of such problems is to combine a well-known numerical methods, such as the Finite Element Method (FEM), in Ω_2 with an enriched analytical solution-based method in Ω_1, see for instance works [5, 9, 16, 20] for examples of existing approaches.

The linear elastic fracture mechanics problem can be formulated as a coupled problem when viewed from the perspective of coupling an elasticity model in Ω_1 with another elasticity model in Ω_2. Within the category theory-based modelling framework, this situation corresponds to the coupling of models from the same category. Specifically, we need to work with the category **Elast** of various elasticity models, see Sect. 3.2.4.2 for details. Thus, according to Definition 3.7, the objects of this category have the following form

$$\mathbf{Set}_{i,i} = \langle \mathbf{Set}_i + \mathbf{Set}_i, \mathbf{Set}_{ii}^{\mathrm{Coup}} \rangle.$$

The structure of these objects clearly indicates that the set of coupling conditions $\mathbf{Set}_{ii}^{\mathrm{Coup}}$ must be formulated. In the fracture mechanics example, these coupling conditions at the coupling interface require:

- continuity of the displacement field, and
- satisfaction of the equilibrium equation for normal stresses.

Consequently, for a consistent coupling in the solution procedure, these conditions must be satisfied when solving the linear elastic fracture mechanics problem shown in Fig. 4.7 in practice.

It is worth emphasising that the category theory-based modelling framework provides a clear perspective on the coupling procedure through an explicit set of coupling conditions. Moreover, the structure of objects $\mathbf{Set}_{i,i}$ is independent of the specific models being coupled. Practically, this means that various models from the category **Elast** can be coupled, and always it will be clear which coupling conditions must be satisfied. Hence, the categorical perspective supports the development of coupled solution procedure by offering a structured way to represent the problem.

Evidently, the category theory-based modelling framework can be applied to develop any coupled solution procedure for problems in mathematical physics. This can be achieved by following the ideas described above for the example from the linear elastic fracture mechanics. We would like to underline once more, that even if a specific coupling realisation is classified as inconsistent in the sense of (4.10), this does not automatically imply that it is incorrect. Highlighting the inconsistency of coupling serves to identify potential sources of errors that might otherwise be overlooked. This is particularly important since the practical realisation of coupling is often treated as a purely algorithmic problem. The category theory-based modelling framework clearly demonstrates that coupling is a conceptual problem.

4.5 Summary of the Chapter

In this chapter, we have illustrated the practical relevance of the category theory-based modelling framework by presenting engineering perspectives on the categorical constructions introduced in Chap. 3 and examples from practical engineering.

To support the practical application of the category theory-based modelling framework, we have introduced simpler analogues of the formal, rigorous definitions in Sect. 4.1. The aim is to capture the essence of formal definitions and abstract constructions, presenting them in a form easily usable by engineers in practice. In this regard, the simplified definitions serve as an interlayer between the abstract constructions and engineering practice. An initial understanding of the category theory-based modelling framework can be achieved by studying the simplified definitions, which can then be complemented by the formal definitions and abstract constructions, if necessary.

In Sect. 4.2, we have discussed model properties such as robustness, sensitivity, and uncertainty, evaluating their potential formalisation within the category theory-based modelling framework. As it has been shown, these model properties are not *universal model properties* but rather *applied model properties*, and, therefore, cannot be properly formalised within the framework. After that, we have introduced the notion of model completeness and the concept of model reduction. Furthermore, the connection between model completeness and model fidelity has been emphasised.

The applicability of the category theory-based modelling framework in real-world engineering problems has been illustrated in Sect. 4.3, with an example from the field of aerodynamic analysis of bridges. This example clearly emphasises how categorical constructions enhance the understanding of relations between different models. Moreover, by working with the relative comparison of models concept proposed in Sect. 4.2, it becomes possible to quantify the influence of specific modelling assumptions on the model response.

Finally, ideas for using the category theory-based modelling framework to develop coupled solution procedures have been presented in Sect. 4.4. In particular, the concept of a *consistent* realisation of model coupling in practice has been introduced. The main idea is that the coupling conditions formulated during the conceptual modelling phase must be appropriately addressed during the instance modelling phase. The categorical perspective provides a structured way to "tracking" the coupling conditions from the conceptual level to practical implementation. Additionally, it has been shown that the development of coupled solution procedures can be formulated in terms of coupling categories of mathematical models. This perspective clearly identifies the coupling conditions that must be satisfied when coupling different solution methods.

References

1. Awodey, S. (2010). *Category theory*. Oxford University Press Inc.
2. Bautechnik, Sonderdruck: GRK 1462 Modellqualitäten 88(6) (2011).
3. Bautechnik, Sonderdruck: GRK 1462 Modellqualitäten 90(4) (2013).
4. Behrisch, M., Kerkhoff, S., Pöschel, R., Schneider, F. M., & Siegmund, S. (2017). Dynamical systems in categories. *Applied Categorical Structures, 15*(1), 29–57.
5. Belytschko, T., Gracie, R., & Ventura, G. (2009). A review of extended/generalized finite element methods for material modelling. *Modelling and Simulation in Materials Science and Engineering, 17*, 043001.
6. Combemale, B., Gray, J., Jézéquel, J.-M., & Rumpe, B. (2024). How does your mode represent the system? A note on model fidelity, underspecification, and uncertainty. *Software and Systems Modeling, 23*, 1053–1054.

7. DFG Research Training Group GRK 1462. (2008–2017). Evaluation of coupled numerical and experimental partial models in structural engineering. Bauhaus-Universität Weimar.
8. Grisvard, P. (1992). *Singularities in boundary value problems*. Springer.
9. Gürlebeck, K., Kähler, U., & Legatiuk, D. (2017). Error estimates for the coupling of analytical and numerical solutions. *Complex Analysis and Operator Theory, 11*(5), 1221–1240.
10. Kavrakov, I., & Morgenthal, G. (2017). A comparative assessment of aerodynamic models for buffeting and flutter of long-span bridges. *Engineering, 3*, 823–838.
11. Kavrakov, I., & Morgenthal, G. (2018). A synergistic study of a CFD and semi-analytical models for aeroelastic analysis of bridges in turbulent wind conditions. *Journal of Fluids and Structures, 82*, 59–85.
12. Kavrakov, I., Legatiuk, D., Gürlebeck, K., & Morgenthal, G. (2019). A categorical perspective towards aerodynamic models for aeroelastic analyses of bridges. *Royal Society Open Science, 6*, 181848.
13. Kavrakov, I., Kareem, A., & Morgenthal, G. (2020). Comparison metrics for time-histories: Application to bridge aerodynamics. *Journal of Engineering Mechanics, 146*(9), 04020093–1.
14. Kavrakov, I. (2020). *Synergistic framework for analysis and model assessment in bridge aerodynamics and aeroelasticity*. PhD Thesis, Bauhaus-Universität Weimar.
15. Lahmer, T., Knabe, T., Nikulla, S., & Reuter, M. (2011). Bewertungsmethoden für Modelle des konstruktiven Ingenieurbaus. *Bautechnik Sonderdruck: GRK 1462 Modellqualitäten, 88*(6), 60–64.
16. Legatiuk, D. (2015). *Evaluation of the coupling between an analytical and a numerical solution for boundary value problems with singularities*. PhD Thesis, Bauhaus-Universität Weimar.
17. Legatiuk, D. (2021). Mathematical modelling by help of category theory: Models and relations between them. *Mathematics, 9*(16), 1946.
18. Liebowitz, H. (1968). *Fracture, an advanced treatise. Volume II: Mathematical fundamentals*. Academic Press.
19. Oberkampf, W. L., & Trucano, T. G. (2002). Verification and validation in computational fluid dynamics. *Progress in Aerospace Sciences, 38*(3), 209–272.
20. Piltner, R. (1985). Special finite elements with holes and internal cracks. *International Journal for Numerical Methods in Engineering, 21*, 1471–1485.
21. Robinson, A. (1963). *Introduction to model theory and to the metamathematics of algebra*. North-Holland Publishing Company.
22. Saltelli, A., Tarantola, S., & Campolongo, F. (2004). *Sensitivity analysis in practice: A guide to assessing scientific models*. John Wiley & Sons Ltd.
23. Spivak, D., & Kent, R. (2012). Ologs: A categorical framework for knowledge representation, *PLoS One, 7*(1), e24274.

Chapter 5
Type-Theoretic and Abstract Algebraic Approaches to Conceptual Modelling

Abstract In this chapter, we explore applications of type theory and abstract algebra to conceptual modelling in engineering. This chapter begins with a presentation of ideas for *automatic model generation*. The key concept is to analyse the formalisation mappings that appear in categories of mathematical models from a computer science perspective. These ideas naturally lead to the application of type theory to mathematical modelling. In this concept, mathematical models are formalised by types, and the type system of a functional programming language can be utilised to verify the consistency of model derivation. A general algorithm for such verification is presented in this chapter. Finally, a relational algebra-based approach to abstract description of models is proposed. While this approach lacks some strictness of the category theory-based modelling framework, it compensates with enhanced flexibility in working with models and relations between them. Moreover, even models of physical objects, which are not necessary described in terms of mathematical expressions and equations, can also be addressed by this approach.

5.1 Ideas Towards Automatic Model Generation

The goal of this section is to initiate a discussion on *automatic model generation*. The idea is to look at the category theory-based modelling framework proposed in Chap. 3 from a computer science perspective. We revisit the central definition of the framework—Definition 3.1, which introduces categories of mathematical models. This revisiting leads naturally to a general discussion on the possibility of the practical implementation of the category theory-based modelling framework. Additionally, we highlight open problems that arise in this context. As a result, this section outlines potential directions for future research.

To better highlight the ideas on automatic model generation, let us introduce a slightly modified version of Definition 3.1:

Definition 5.1 Let $\mathbf{Model}_1$ be a category of mathematical models describing a given physical phenomenon. Additionally, we assume that there exists a finite set

D. Legatiuk, *Mathematical Modelling by Help of Category Theory*,
Mathematical Engineering, https://doi.org/10.1007/978-3-031-84951-0_5

X containing all possible assumptions which are needed to build the models. Then the category $\mathbf{Model}_1$ has the following structure:

1. each object is a subset of the set X—*set of assumptions* of a mathematical model, denoted by $\mathbf{Set}_{\mathrm{A}}$, where A is the corresponding mathematical model;
2. and there is a morphism from $\mathbf{Set}_{\mathrm{A}}$ to $\mathbf{Set}_{\mathrm{B}}$, precisely then $\mathbf{Set}_{\mathrm{B}} \subset \mathbf{Set}_{\mathrm{A}}$;
3. for each set of assumptions and its corresponding model exists a mapping
$$\mathbf{Set}_{\mathrm{A}} \overset{S}{\mapsto} \mathrm{A};$$
4. all objects are related to mathematical models acting in the same physical dimension.

This definition is, in a sense, more natural from the perspective of category theory, because it does not require the explicit construction of objects, in contrast to Definition 3.1, where it is implicitly assumed, that modelling assumptions have to be grouped into sets. Definition 5.1 only assumes the existence of a set X containing all modelling assumptions relevant to the consideration of a specific phenomenon.

From a broader perspective, Definition 5.1 can be seen as a step towards an *automatic model generation*. Given a set X, such as for example a database of modelling assumptions, it is "only" necessary to describe how specific sets of assumptions are constructed and subsequently formalised. To elaborate on this point, we provide an idea of how sets of assumptions can be constructed in an abstract manner, without requiring a person to "review" their validity. As we will see, this approach raises additional fundamental questions that, while beyond the scope of the current work, are nonetheless worth mentioning.

We now present a general idea for constructing categories of mathematical models without using prior knowledge of the structure of sets of assumptions for specific models. We will briefly outline the steps constituting this idea, followed by a discussion:

1. First, a set X of all possible model assumptions to be used in a category **Model** is created.
2. Second, the power set of X, $\mathcal{P}(X)$, is constructed.
3. Third, a formalisation functor $S\colon \mathbf{Model} \to \mathbf{Model}^F$ is applied to each element of $\mathcal{P}(X)$.
4. Fourth, subsets of $\mathcal{P}(X)$, which are *consistently formalised* by functor S, are selected as objects of **Model**. Thus, a category of mathematical models is created.

Steps 1 and 2 are straightforward, because the only requirement for these steps is the existence of a set X containing all possible modelling assumptions, which, as it has been mentioned earlier, can be provided in the form of a database. However, Steps 3 and 4 are significantly more challenging and require a more detailed discussion. First, recall that the formalisation mapping S is not unique, and generally not invertible, see the discussion on Definition 3.1. Additionally, note that technically, at Step 3, the category **Model** still has no objects, because not all elements of $\mathcal{P}(X)$ contain a set of assumptions corresponding to a mathematical model.

The key point to the success of Steps 3 and 4 is a formal definition of a *consistently formalisable set of assumptions*. As we will see, this question quickly leads to a fundamental problem at the intersection of mathematics and computer science. The issue arises because evidently not all elements of $\mathcal{P}(X)$ are sets of assumptions that lead to meaningful expressions representing mathematical models. Hence, it becomes necessary to establish criteria for deciding whether a subset of $\mathcal{P}(X)$ must be kept as an object of category **Model** or discarded. This problem can be seen as an analogue to the *decidability problem* in logic, see for example [11], or to the *decision problem* in computer science, see [18] among others. To make the analogy explicit, let us recall the decidability problem from [11]:

> A system S is said to be *decidable* if and only if there is an effective procedure whereby, for any given well-formed formula α, it can be determined in a finite number of steps whether or not α is a theorem of S.

There are several formulations of the decision problem discussed in [18]. For the purpose of our discussion, we recall the formulation of Hilbert's decision problem:

> Let $\mathcal{L}$ be a first-order language. Find an algorithm that, given an arbitrary formula A of $\mathcal{L}$, decides (YES or NO) whether A is logically valid (that is, true in every interpretation of $\mathcal{L}$). By Gödel's Completeness Theorem, this is equivalent to asking if A is a theorem of first-order logic.

Because of the fundamental nature of the problem, it is difficult to properly define a consistently formalisable set of assumptions. Nonetheless, as an initial idea in this direction, we propose the following formulation:

Definition 5.2 (Consistently Formalisable Set of Assumptions) Let S be a formalisation mapping, and let $\mathbf{Set}_\mathrm{A}$ be a set of assumptions written by help of a natural language. We call the set $\mathbf{Set}_\mathrm{A}$ a *consistently formalisable set of assumptions*, if and only if action of S on $\mathbf{Set}_\mathrm{A}$ leads to a valid model formulation, and there is a possibility, at least theoretical, to verify the derivation of this model formulation from $\mathbf{Set}_\mathrm{A}$.

As formulated, this definition makes clear the connection between the verification of model derivation and decidability/decision problems discussed above. Additionally, the idea of Definition 5.2 is to propose a criterion, at least theoretical, to avoid "pathological models", which are, on the one hand, obtained by formalising some elements of $\mathcal{P}(X)$, but, on the other hand, do not have any practical meaning.

To further elaborate the point of having pathological models, let us recall the category **Beam** containing sets of assumptions of beam theories listed in Table 3.1. In this case, the set X of all modelling assumptions used in **Beam** is given by:

$$X = \left\{ \begin{array}{l} \text{1. Cross sections of a beam that are planes remain planes after} \\ \quad \text{the deformation process} \\ \text{2. Normal stresses on planes parallel to the axis of a beam} \\ \quad \text{are infinitesimal} \\ \text{3. A beam has a constant cross section} \\ \text{4. A beam is made of a homogeneous isotropic material} \\ \text{5. Cross sections of a beam perpendicular to its axis remain} \\ \quad \text{perpendicular to the deformed axis} \\ \text{6. Rotation inertia of cross sections of a beam is omitted} \end{array} \right\}$$

Thus, $\mathcal{P}(X)$ contains 64 elements from which only three, namely

$$\{1, 2, 3, 4, 5, 6\}, \quad \{1, 2, 3, 4, 5\}, \quad \{1, 2, 3, 4\},$$

lead to valid model formulations. Hence, these three sets are consistently formalisable sets of assumptions. The remaining 61 elements of $\mathcal{P}(X)$ are not consistently formalisable, and, thus, must be removed from the consideration.

Definition 5.2 also necessitates a verification procedure to sieve out only the consistently formalisable sets of assumptions from $\mathcal{P}(X)$. In the next section, we present first ideas based on functional programming and type theory to address this verification procedure. However, it is necessary to emphasise, that these results are not yet complete and should be regarded as a preliminary step towards the practical implementation of the category theory-based modelling framework.

We conclude this section with the following remark:

Remark 5.1 From the perspective of theoretical computer science, a verification procedure for the formalisation of modelling assumptions requires a formal translation from natural language into mathematical expressions. The general question of the existence of such a formal translation and its rigour are open questions that extend beyond the scope of this work. Moreover, given that the set of all modelling assumptions is still finite, it is possible to overcome this problem by enforcing stricter rules on the formulation of sets of modelling assumptions.

5.2 Type-Theoretic Approach to Conceptual Modelling

In this section, we present ideas of using type theory to formalise mathematical models, see [26] for a general information on type theory. The overall goal is to propose an algorithm that allows verification of model derivation, as discussed in the previous section. In works [10, 16], this verification of model derivation has been referred to as a *model coherence check* or *formal detection of modelling errors*. Although the category theory-based modelling framework has not been directly addressed in these works, the mathematical models in this context were, in fact, model formulations derived through the formalisation of sets of assumptions.

In addition to focusing on the category $\mathbf{Model}^F$, the category of model formulations, the works [10, 16] have also addressed the influence of modelling assumptions on the modification of models. This means that the formalisation process of sets of assumptions, and consequently a verification strategy for checking the consistency of this formalisation process have also been considered. In this section, we will present the ideas from [10, 16], adapting the terminology of these works to the current context.

A formal verification procedure for checking the consistency of model derivation introduced requires the development of a framework that can be implemented on a computer. Possible formal foundations for such a framework can generally be divided into two main groups:

- The first possibility is to use automated theorem proving as a basis. In this case, the corresponding mathematical model must be written in terms of logical expressions, see for example [30] for details. In this setting, verifying a model derivation would require checking the deducability of the final logical expression.
- The second possibility is to use type theory. In this case, a particular type is assigned to each basic model component, and a complete model is constructed by combining these basic model components through matching their types. Verifying a model derivation is then performed on the level of a type system governing the conditions under which typed components can be combined, inferring the type of the combined entity from this.

Furthermore, it is worth mentioning, other formal foundations, such as for example *dependent types* which provide possibilities to express arbitrary logical predicated at the type level, see for example [1]. Some "mixed" systems have been already implemented, for example LiquidHaskell [12] demonstrates one approach to support selected classes of predicates at the type level while retaining full automation through the integration of a satisfiability modulo theories solver (SMT solver, see [14] for details). The choice of a specific formal foundation is, in fact, depends on desired expressiveness of the resulting system, as well as its strictness. Because evidently, than more strictness we impose, then less expressiveness we obtain.

Another advantage of using type theory for verification of model derivation is that many programming languages, especially functional programming languages such as Haskell [2], utilise the idea of composing types. Hence, the type checker of Haskell can be used for the verification process. Thus, the ideas presented in this section can be implemented as a Haskell embedding. But this is the task of future work, because such a Haskell embedding would also require certain support of symbolic computations, which is not current possible, see [10] for details on this point.

It is important to recall the FHM (Functional Hybrid Modelling), which is a prototype Haskell-embedded modelling language [23, 24] supporting mathematical modelling in terms of differential-algebraic equations. FHM is an example of how type checker of Haskell can be used for model consistency check. In particular, supported by formal semantic [4–6], FHM has been used to ensure solvability of the system of equations, which is done using the Haskell type checker via appropriately constructed structural types.

5.2.1 Signals and Signal Relations

This section presents basic ideas of extending FHM to mathematical modelling with partial differential equations. The basic "unit" of FHM is a *signal*, which is a type-varying value—a function from time to a value of a specific type. Let us now make the following remark:

Remark 5.2 The term "signal" was a natural choice in the FHM, since it has been designed for initial value problems represented by differential-algebraic equations with applications in electrical engineering. Later on, ideas of FHM have been extended to address also problems in mathematical physics, which are represented by partial differential equations, see again [10, 16]. Nonetheless, the usage of the term *signal* for the unknowns has been preserved, but it must be kept in mind that signals are now more general as they can be *temporal*, *spatial*, *spatio-temporal*, or even *constant*. The distinction between different kinds of signals should become clear from the signal type signature.

To establish a type system applicable to different problems of mathematical physics, we first introduce the finite-dimensional set $\{\xi_i\}_{i=0}^n$ of variables with the conventions that ξ_0 corresponds to the time variable, and ξ_i, $i = 1, \ldots, n$, correspond to the spatial variables used in a model, and n being the total number of spatial dimensions. For simplicity, we take the type of the time variable and the spatial variables to be $\mathbb{R}$, and to avoid confusion about the ordering of ξ_i, we assume that the ordering of the spatial variables follows the standard ordering of unit vectors in $\mathbb{R}^n$. It follows that, for now, we make the simplifying assumption that all models in principle are time-dependent, while the number of spatial dimensions can vary.

Based on the above discussion, the first step in the verification of model derivation is reformulating a model formulation in terms of the new variables ξ_i, $i = 0, 1, \ldots, n$. The reformulation allows us to introduce the polymorphic type of signals as follows:

$$\mathsf{S}\,\alpha = \mathbb{R}^{(n+1)} \rightarrow \alpha,$$

where $\mathbb{R}^{(n+1)}$ is the type of a vector of time and space coordinates, and α is the type of values "carried" by the signal; i.e., its value at a specific point in time and space. The concrete type α depends on the model under consideration: for scalar-valued quantities, α is a base type; for vector-valued quantities, α is a product type.

A signal thus represents the unknown function in the equation(s) of a model formulation. Signals therefore only exist implicitly: otherwise the solution of a problem would already be given from the beginning. Consequently, the signal type is *conceptual* and never used explicitly. What is typed explicitly are the *equations* that characterise signals, leading us to the notion of *signal relations* to which we will turn shortly.

For the convenience of future practical realisation of the language proposed in this section, and to express n-ary relations on signals, we view a signal carrying elements of a product type

$$\mathsf{S}\,(T_1 \times T_2 \times \cdots \times T_n)$$

as isomorphic to a product of signals

$$(\mathsf{S}\,T_1) \times (\mathsf{S}\,T_2) \times \cdots \times (\mathsf{S}\,T_n)\,.$$

It is also important to make the following remark:

Remark 5.3 In practice, following the variable convention ξ_i, $i = 0, 1, \ldots, n$ set out above means that a problem formulated in tensor setting [19], complex setting [21], or hypercomplex setting [8] needs to be re-written in a component form. It may seem restrictive to put strong demands on the formulation of models, but after completing the type-theoretic setting for componentwise formulations of problems, the extension of basic type system to more general formulations will be straightforward.

We now introduce the type of *signal relations*: relations on signals of some specific spatial dimensionality n. The type

$$\mathsf{SR}\,(n :: \mathbb{N})\,\alpha$$

is the type for a relation on a signal of type $\mathsf{S}\,\alpha$ depending on time and n spatial dimensions with $\mathbb{N} = \{0, 1, 2, \ldots\}$ denoting the set of natural numbers. To relate two or more signals, we rely on the isomorphism between a product of signals and a signal of products, as discussed above. For example, the type of a relation between two signals of n spatial dimensions carrying concrete types T_1 and T_2, respectively, would be $\mathsf{SR}\,n\,(T_1, T_2)$.

For the general notations for defining relations on signals, we use the following notation originating in λ-abstraction

$$\textbf{sigrel} \quad \mathit{pattern} \quad \textbf{where} \quad \mathit{equations}$$

The *pattern* introduces signal variables that are bound to the value of the corresponding signal at each point in time and space. For example, for a given signal variable p of a given type t, i.e. $p :: t$, the notation introduced above leads to

$$\textbf{sigrel} \quad p \quad \textbf{where} \ldots :: \mathsf{SR}\,n\,t$$

To describe concrete model formulations we introduce two kinds of equations:

$$e_1 = e_2, \qquad sr \ \diamond e_3, \tag{5.1}$$

where e_i, $i = 1, 2, 3$ are expressions that are allowed to introduce new variables and sr is an expression denoting a signal relation.

The signals characterised by Eqs. (5.1) and all involved signal relations are all for the same number n of spatial dimensions, as reflected by the type of the overall signal relation. Both equations are further restricted to be well-typed: if $e_i :: T_i$, $i = 1, 2, 3$, then $T_1 = T_2$ and $sr :: \mathsf{SR}\, n\, T_3$. The first kind of equation requires the values of the two expressions to be equal at all points in space and time. The second kind of equation introduces arbitrary relations on signals, where the symbol $\diamond$ is understood as a *relation application* resulting in a constraint that must hold at all points in space and time. Moreover, naturally the first kind of equation is just a special case of the second kind, as equality is a subset of general relations on signals.

Looking at practical applications of category theory-based modelling framework, it might be necessary to couple mathematical models with different number of spatial dimensions. Therefore, it is necessary to address models with different number of spatial dimensions at the level of types and signals. This can be accomplished by an operation that transforms a signal relation by applying an affine transformation to the (spatial part of) the coordinate vectors. Such a transformation might have a type along the lines

$$\mathsf{SR}\, m\, \alpha \rightarrow \mathbb{R}^{m \times (n+1)} \rightarrow \mathsf{SR}\, n\, \alpha$$

where $\mathbb{R}^{m \times (n+1)}$ is the type of an m by $n + 1$ real-valued matrix representing the affine transformation of coordinates in the space of the resulting signal relation of type $\mathsf{SR}\, n\, \alpha$ (n spatial dimensions) to coordinates in the space of the transformed signal relation of type $\mathsf{SR}\, m\, \alpha$ (m spatial dimensions).

As we do not need any such operation in the following, we do not discuss this point further here. However, we do note that the above type is an example of a *dependent* type [1], of which we will encounter further examples in the following. In more detail, the actual type of the transformation would be:

$$\{m\, n :: \mathbb{N}\} \rightarrow \mathsf{SR}\, m\, \alpha \rightarrow \mathbb{R}^{m \times (n+1)} \rightarrow \mathsf{SR}\, n\, \alpha$$

Thus, the dimensions of the signal relations and the size of the transformation matrix depends on the *values* of the first two arguments m and n to the transformation operation; hence *dependent* type. Following the conventions of the dependently typed language Agda [25], the two first arguments have been enclosed in braces to indicate that they are *implicit*, meaning that they can be left out whenever they can be inferred from the context, which usually is the case for the number of dimensions of a model.

5.2.2 *Typing of Few Basic Operators*

The practical use of signal relations also requires typing of the mathematical tools used to derive a model formulation via a formalisation mapping S. In this section, we discuss this typing on the example of heat equation:

$$c(\mathbf{x})\rho(\mathbf{x})\frac{\partial\theta}{\partial t} - \operatorname{div}(\lambda(\mathbf{x})\operatorname{grad}\theta) = f(\mathbf{x}, t). \tag{5.2}$$

The mathematical operators, such as ∂_t, div, grad, as well as the unknown signal $\theta(\mathbf{x}, t)$, the heat capacity $c(\mathbf{x})$, the material density $\rho(\mathbf{x})$, and the variable thermal conductivity $\lambda(\mathbf{x})$ must be typed. For the sake of completeness, we recall the definitions of divergence and gradient:

$$\operatorname{div}\mathbf{u} := \nabla\cdot\mathbf{u} = \frac{\partial u_1}{\partial x_1} + \frac{\partial u_2}{\partial x_2} + \cdots + \frac{\partial u_n}{\partial x_n},$$

$$\operatorname{grad} f := \nabla f = \frac{\partial f}{\partial x_1}\mathbf{e}_1 + \frac{\partial f}{\partial x_2}\mathbf{e}_2 + \cdots + \frac{\partial f}{\partial x_n}\mathbf{e}_n,$$

where $\mathbf{u} = (u_1, u_2, \ldots, u_n)$ is a vector-valued function of n variables x_i, $i = 1, \ldots, n$, $f = f(x_1, \ldots, x_n)$ is a scalar-valued function, and $\mathbf{e}_i$, $i = 1, \ldots, n$ are the standard unit vectors in $\mathbb{R}^n$.

Assuming the convention for new variables discussed above, $\{\xi_i\}_{i=0}^n$, the partial differentiation with respect to individual variables can be written as a general operator [16]:

$$D_i^s := \frac{\partial^s}{\partial \xi_i^s}, \tag{5.3}$$

where i is the index of a differentiation variable and s is the order of the derivative. The operator has to satisfy the standard laws:

$$D_i^n D_i^m = D_i^m D_i^n = D_i^{n+m}, \quad D_i^n D_j^m = D_j^m D_i^n, \quad D_i^0 = I, \tag{5.4}$$

where I is the identity operator. These rules have to be reflected in the type system.

The type of the operator D_i^s is

$$D :: \{n :: \mathbb{N}\} \to \mathbb{N}_{\leq n} \to \mathbb{N} \to \mathsf{SR}\, n\, (\alpha, \alpha), \tag{5.5}$$

where $\mathbb{N}_{\leq n}$ is the type of natural numbers up to and including n. Note that this is another example of a dependent type, with the dimension argument implicit as it usually will be implied by the context. Both the variable index i and the derivative order s are restricted to natural numbers as we do not consider fractional derivatives at present.

As divergence and gradient require only first order partial derivatives with respect to the *spatial* variables, the divergence and gradient can be defined by a recursive application of operator D_i^1 to a vector-valued signal and via mapping the operator D_i^1 over a vector filled in by a scalar-valued function, respectively. The types of div and grad are:

$$\begin{aligned} \texttt{div} &:: \{n :: \texttt{Nat}\} \rightarrow \mathrm{SR}\,n\,(\alpha^n, \alpha), \\ \texttt{grad} &:: \{n :: \texttt{Nat}\} \rightarrow \mathrm{SR}\,n\,(\alpha, \alpha^n), \end{aligned} \tag{5.6}$$

where α is a base type and α^n is an n-ary vector of α; i.e., a product of n αs. Note that this, again, is a dependent type, and that the number of spatial dimensions, again, is an implicit argument, allowing it to be omitted whenever it is implied by the context, which we exploit in the following.

The composition of these operators is not commutative:

$$\texttt{div} \circ \texttt{grad} :: \mathrm{SR}\,n\,(\alpha, \alpha), \text{ while } \texttt{grad} \circ \texttt{div} :: \mathrm{SR}\,n\,(\alpha^n, \alpha^n),$$

where $\circ$ is relational composition. Moreover, $\texttt{div} \circ \texttt{grad}$ defines the classical Laplace operator, which can be applied to scalar-valued functions or mapped over vector-valued functions.

The operators D, `div`, and `grad` can now be used to relate signals by means of relation application:

$$D(i)(s) \diamond (f_1, g_1), \quad \texttt{div} \diamond (f_2, g_2), \quad \texttt{grad} \diamond (f_3, g_3)$$

The meaning of this is given by the following equations:

$$f_1(\boldsymbol{\xi}) = \frac{\partial^s g_1(\boldsymbol{\xi})}{\partial \xi_i^s}, \quad f_2(\boldsymbol{\xi}) = \sum_{j=1}^{n} \frac{\partial (g_3)_j(\boldsymbol{\xi})}{\partial \xi_j}, \quad f_3(\boldsymbol{\xi}) = \sum_{j=1}^{n} \frac{\partial g_3(\boldsymbol{\xi})}{\partial \xi_j} \mathbf{e}_j$$

where $\boldsymbol{\xi}$ denotes the vector of the variables $\xi_0, \xi_1, \ldots, \xi_n$.

To make it easier to write equations, a surface syntax, closer to the classical mathematical notations than the relational syntax introduced above, could be adopted:

$$f_1 = \mathrm{D}(i)(s)\, g_1, \quad f_2 = \texttt{div}\, g_2, \quad f_3 = \texttt{grad}\, g_3.$$

With this notation, we can view application of D, `div` and `grad` to a signal as resulting in a new signal. This is just a more concise, arguably more intuitive, syntax: the underlying meaning is exactly as above.

5.2.2.1 Example of the Rayleigh Model

Let us illustrate the proposed language on the example of the Rayleigh beam model, which is a formalisation of the object $\mathbf{Set}_{\mathrm{R}}$ in the category **Beam**, see Sect. 3.2.4.1 for the categorical perspective. Let us recall the form of this model formulation:

$$\mathbf{Set}_{\mathrm{R}} \overset{S}{\mapsto} \rho\, F \frac{\partial^2 u}{\partial t^2} + E\, I_y \frac{\partial^4 u}{\partial x^4} - \rho\, I_y \frac{\partial^4 u}{\partial x^2 \partial t^2} = 0$$

where S is a formalisation mapping. By help of the new convention with general variables $\{\xi_i\}_{i=0}^n$, the differential equation of the Rayleigh beam theory is written as follows:

$$\rho F \frac{\partial^2 u}{\partial \xi_0^2} + E I_y \frac{\partial^4 u}{\partial \xi_1^4} - \rho I_y \frac{\partial^4 u}{\partial \xi_1^2 \partial \xi_0^2} = q(x, t). \tag{5.7}$$

Using signals and signal relations, model formulation of $\mathbf{Set}_{\mathrm{R}}$ can be now defined as follows:

$$\begin{aligned} &Rayleigh :: \mathsf{SR}\,(\mathbb{R}, \mathbb{R}) \\ &Rayleigh = \mathbf{sigrel} \quad (u, q) \quad \mathbf{where} \\ &\qquad q = \rho \cdot F \cdot D(0)(2)u + E \cdot Iy \cdot D(1)(4)u \\ &\qquad\quad -\rho \cdot Iy \cdot (D(1)(2) \circ D(0)(2))\, u \end{aligned}$$

where $D(i)(n) \circ D(j)(m)$ denotes the composition of two operators D obeying rules (5.4), the material constants ρ, F, E, Iy are understood as signals which are constant at all points in space and time and the equality relation represent the principle meaning of any mathematical physics model given by an equation.

For illustrative purposes, let us also show this example expressed directly using signal relations and application, which is the underlying representation:

$$\begin{aligned} &Rayleigh :: \mathsf{SR}\,(\mathbb{R}, \mathbb{R}) \\ &Rayleigh = \mathbf{sigrel} \quad (u, q) \quad \mathbf{where} \\ &\qquad q = \rho \cdot F \cdot d2dt2u + E \cdot Iy \cdot d4dx4u \\ &\qquad\quad -\rho \cdot Iy \cdot d4dx2dt2u \\ &\qquad D(0)(2) \qquad\qquad\quad \diamond\ (u, d2dt2u) \\ &\qquad D(1)(4) \qquad\qquad\quad \diamond\ (u, d4dx4u) \\ &\qquad D(1)(2) \circ D(0)(2)\ \diamond\ (u, d4dx2dt2u) \end{aligned}$$

5.2.3 General Algorithm for Verifying a Model Formulation Derivation

We now outline an algorithm for the verification of model formulation derivation using the heat equation (5.2) as an example. The inconsistencies in model derivation can be detected by checking the types of basic modelling assumptions constituting the model of heat conduction that ultimately results in Eq. (5.2), as opposed to checking the final model formulation only.

Basic modelling assumptions for a heat conduction process have been already listed in Table 3.3 while discussing the category **Heat**. For illustrative purposes, we recall assumptions related to the heat equation (5.2) and discuss how these assumptions are formalised:

1. Fourier's law: if the temperature of a body is non-uniform, heat currents arise in the body, directed from points of higher temperature to points of lower temperature. For the heat flux density $\mathbf{q}$ this reads as

$$\mathbf{q} = -\lambda \operatorname{grad} u. \tag{5.8}$$

   Integrating (5.8) over the surface of an arbitrary reference domain $\Omega_0 \subset \Omega$ we obtain

$$Q_1 = -dt \int_{\partial\Omega_0} \lambda(x) \frac{\partial u}{\partial n} d\Gamma, \tag{5.9}$$

   where Q_1 is the amount of heat flowing through the surface of Ω_0 in the time interval of length dt and $d\Gamma$ stands for the area element on $\partial\Omega_0$.
2. The amount of heat that is needed for changing the temperature of Ω_0 in the time interval of length dt is equal to

$$Q_2 = dt \int_{\Omega_0} c\rho \frac{\partial u(x,t)}{\partial t} dV. \tag{5.10}$$

3. The amount of heat generated or absorbed inside the body is given by

$$Q_3 = dt \int_{\Omega_0} F(x,t) dV, \tag{5.11}$$

   where $F(x,t)$ is the density of the heat source or the heat sink, and dV the volume element.

Using formulae (5.9)–(5.11) and the principle of conservation of energy we get:

$$\int_{\Omega_0} c\rho \frac{\partial u(x,t)}{\partial t} dV = \int_{\partial\Omega_0} \lambda(x) \frac{\partial u}{\partial n} d\Gamma + \int_{\Omega_0} F(x,t) dV. \tag{5.12}$$

Additionally, we need to introduce two mathematical relations, playing the role of basic operators. One is the Gauß integral theorem, and the other is the concept of defining a density by shrinking Ω_0 to a point x by taking the limit (under certain assumptions for f)

$$\lim_{\Omega_0 \mapsto \{x\}} \frac{\int_{\Omega_0} f(s) dV}{\int_{\Omega_0} 1 dV} = f(x).$$

Finally, applying these two mathematical tools to integral (5.12) yields the differential equation (5.2).

In our setting, the definite integral in one dimension can be typed straightforwardly as follows:

$$\texttt{integral} :: \{n :: \mathbb{N}\} \to \mathbb{N}_{\leq n} \to (\mathbb{R}, \mathbb{R}) \to \mathsf{SR}\, n\, (\alpha, \alpha), \tag{5.13}$$

where $\mathbb{N}_{\leq n}$ is the type of the index of the integration variable and $(\mathbb{R}, \mathbb{R})$ is the type of the integral bounds. Multi-dimensional integration can now be defined by iterative application of (5.13).

Using the integral above together with the notation and operators defined in the previous section, the basic modelling assumptions (5.8) to (5.12) can now be transliterated into our typed setting. However, as the objective of this section is to illustrate that a verification procedure for model derivation is indeed possible, we omit the details of this transliteration. Instead, we discuss how the typed modelling framework outlined above can be used as a foundation for an algorithm for checking model derivation:

Given:	Typing of basic modelling assumptions in a form of a library.
`Step 1:`	Declare concrete types for variables, functions, coefficients;
`Step 2:`	Select basic modelling assumptions from the library;
`Step 3:`	Compiler checks if the resulting expression is well-typed;
`Step 4:`	Performing symbolic calculations on the model equations, if possible;
Output:	Final form of the model with type signatures of all its components.

The algorithm has two main assumptions:

1. the basic modelling assumptions are formalised in our typed framework;
2. the compiler provides support for symbolic calculations.

The first assumption has been addressed earlier when introducing typing of basic mathematical operators. The mathematical operators are considered as primitive objects of formalisation of modelling assumptions. Thus, basic modelling assumptions can be introduced as compositions of basic operators, for example

$$\texttt{Fourier's\_law} := -S \cdot \texttt{integral}\,0\,(t1, t2) \circ (\lambda \cdot D(1)(1))\,,$$

and the Haskell type checker will infer the types of the modelling assumptions. In this way, the library of typed basic modelling assumptions can be created.

The second assumption on the algorithm is mostly a question of implementation using suitable components such as a host language compiler and a computer algebra system. Consider a formal representation of the model of heat conduction as a composition of basic modelling assumptions:

$$\begin{aligned}\texttt{Heat\_Model} := \texttt{Diff}\ \circ (&\texttt{Conservation\_energy\_principle} \circ \\ &\circ \big(\texttt{Fourier's\_law} \circ \texttt{Amount\_of\_heat} \circ \texttt{Heat\_source}\big)\big)\,,\end{aligned}$$

where `Fourier's_law`, `Amount_of_heat`, and `Heat_source` are formalised modelling assumptions, `Conservation_energy_principle` is the formalised principle of conversation of energy leading to the integral formulation (5.12), `Diff` is the function converting the integral form of the heat equation into differential form (5.2) by using the Gauß integral theorem and taking the limit. We use parentheses to avoid making assumptions regarding the associativity of composition.

An alternative, simplified, approach to the typing of basic physical assumptions and providing symbolic calculations capabilities, would be to declare directly the differential form of the model of heat conduction:

$$\begin{aligned}&Heat :: \mathrm{SR}\,3\,(\mathbb{R}, \mathbb{R})\\ &Heat = \textbf{sigrel}\quad (\theta, f)\quad \textbf{where}\\ &\qquad f = c \cdot \rho \cdot D(0)(1)\,\theta - \texttt{div} \circ (\lambda \cdot \texttt{grad})\,\theta\\ &\qquad \lambda = expression\end{aligned}$$

where material constants c and ρ are understood as constant signals, while the thermal conductivity λ is a signal, which is not necessarily a constant. This alternative approach is less general than the first approach, but perhaps more transparent. Moreover, the support by symbolic calculations is also beneficial: if the expression for λ is indeed a constant, the whole expression can be simplified. We would like to remark, that the question of regularity constrains of functions and parameters is not in the scope of current research, although it is important for practical use of models.

Independent on the alternative chosen for type-theoretic model formulation representation, the proposed type-theoretic formalism allows verification of a model derivation discussed in Sect. 5.1. Because the final expression must pass the Haskell type checker, any expression, i.e. model formulation, which is not well-typed will be rejected as inconsistent. Thus, the results presented in this section shows, that the concept of consistently formalisable sets of assumptions introduced in Definition 5.2 is indeed relevant for practical applications, as well as that verification of derivation of model formulations is possible, but requiring deeper studies in theoretical computer science.

We conclude this section with the following remarks:

Remark 5.4 The type-theoretic approach presented in this section can also be extended to detect instance modelling errors appearing in practice, such as for example unit errors. In [7], so-called *dimensional types* have been introduced as an enrichment of the type-theoretic ideas presented in this section. Moreover, this approach has also been implemented in F# programming language, indicating its practical applicability for catching unit errors during the instance modelling phase.

Remark 5.5 Finally, we would like to emphasise once again that the ideas presented in Sects. 5.1–5.2 are preliminary steps towards automatic model generation and the development of a verification procedure for model derivations. These results are not yet fully developed and are intended to serve as a basis for future research and discussion.

5.3 Consistency of Coupling

The ideas presented in Sects. 5.1–5.2 naturally lead to extending the concept of consistency of coupling, as discussed in Sect. 4.4, to the conceptual level. In this context, it is necessary to work at the level of models, meaning that the categories of formalised sets of assumptions $\mathbf{Model}_i^F$, $\mathbf{Model}_j^F$, $\mathbf{ModCoup}_{i,j}^F$, and $\mathbf{ModCoup}_{i,i}^F$ are considered.

We summarise now the discussion in the form of the following proposition, which is a refined version of the result presented in [9]:

Proposition 5.1 *Let us consider the categories* $\mathbf{Model}_i^F$, $\mathbf{Model}_j^F$, *and* $\mathbf{ModCoup}_{i,j}^F$. *Further, let models (objects) in* $\mathbf{Model}_i^F$ *are formalised by a functor* S_1, *models in* $\mathbf{Model}_j^F$ *are formalised by a functor* S_2, *and coupled models in* $\mathbf{ModCoup}_{i,j}^F$ *are formalised by a functor* S_3. *Next, let* $\mathbf{T}\colon \mathbf{Model}_i^F \to \mathbf{ModCoup}_{i,j}^F$ *and* $\mathbf{F}\colon \mathbf{Model}_j^F \to \mathbf{ModCoup}_{i,j}^F$ *be a pair of functors mapping single models into their coupled counterparts. Then the coupling of models from categories* $\mathbf{Model}_i^F$

and $\mathbf{Model}_j^F$ *is said to be consistent if and only if the two squares in the following diagram commute*

$$\begin{array}{ccccc}
\mathrm{A}_1 & \xrightarrow{\mathbf{T}} & \mathrm{A}_1 \times \mathrm{B}_1 & \xleftarrow{\mathbf{F}} & \mathrm{B}_1 \\
\Big\downarrow{\scriptstyle S_1(f)} & & \Big\downarrow{\scriptstyle S_3(s)} & & \Big\downarrow{\scriptstyle S_2(g)} \\
\mathrm{A}_2 & \xrightarrow{\mathbf{T}} & \mathrm{A}_2 \times \mathrm{B}_2 & \xleftarrow{\mathbf{F}} & \mathrm{B}_2
\end{array}$$

for any pairs of objects $\mathrm{A}_1, \mathrm{A}_2 \in \mathbf{Model}_i^F$, $\mathrm{B}_1, \mathrm{B}_2 \in \mathbf{Model}_j^F$, *and* $\mathrm{A}_1 \times \mathrm{B}_1, \mathrm{A}_2 \times \mathrm{B}_2 \in \mathbf{ModCoup}_{i,j}^F$, *and where* f, g, s *are morphisms in categories* $\mathbf{Model}_i$, $\mathbf{Model}_j$, *and* $\mathbf{ModCoup}_{i,j}$, *respectively.*

The idea of this proposition is that the formalisation of modelling assumptions must be performed correctly regardless of whether it applies to individual models or directly to a coupled model. Consequently, if the formalisation has not been done correctly, then the above diagram would not commute, and such a coupling must be rejected as invalid.

Similar to the discussion in Sect. 4.4, the above proposition can be seen as kind of guidance or reference for further development of the type theory-based modelling language. The algorithm for checking model derivations, presented in the previous section, must also be applicable for coupled models, and, thus, consistency of coupling would be automatically verified.

5.4 Alternative View on Mathematical Models

In this section, we present an alternative perspective on mathematical models by relaxing the structural constrains of the categorical constructions presented in Chap. 3. The main motivation is to offer a more flexible, but still mathematically sound, approach to the conceptual modelling of engineering systems. As we will see, this flexible approach allows also consideration of more general models, than only mathematical models.

5.4.1 Mathematical Models as Abstract Triples

As a summary of all ideas on mathematical models, we can now introduce the following abstract definition of a mathematical model:

Definition 5.3 A mathematical model $\mathfrak{M}$ is a triple $\mathfrak{M} = \langle \mathbf{Set}, \mathcal{M}, S \rangle$, where

- **Set** is the set of assumptions of the model;
- $\mathcal{M}$ is an instantiation of the model in terms of mathematical expressions and equations;
- S is a formalisation mapping, which formalises the set of assumptions **Set** into the model instantiation $\mathcal{M}$.

Evidently, this definition encompasses everything that has been discussed about mathematical models: models are not just one specific object, but rather a collection of several connected objects of different nature. Furthermore, relations between the mathematical models can once again be introduced using Definition 3.2.

Additionally, Definition 5.3 proposes an abstract description of a mathematical model, which does not require a formal categorical structure, but, in fact, would not be possible to introduce without all previous results discussed in this book. It is worth emphasising that, while Definition 5.3 provides an abstract description of a mathematical model, it is still more *concrete* than the categorical description. This is because it explicitly requires naming a model instantiation and a formalisation mapping, whereas the categorical approach keeps these objects abstract under functorial mappings and their natural transformations.

We will next revisit some of the examples discussed previously in Chap. 3 for analysing the perspective provided by Definition 5.3, as well as for establishing a clear link to the category theory-based description of mathematical models.

Let us consider again models of beam theories. According to Definition 5.3, these models are not covered any more by a common ground, meaning that they are not explicitly associated to objects in any category. Moreover, it is necessary to explicitly describe these mathematical models as triples:

- A mathematical model of the Bernoulli-Euler beam theory is the triple

$$\mathfrak{M}_{\mathrm{B-E}} := \langle \mathbf{Set}_{\mathrm{B-E}}, \mathcal{A}, S_1 \rangle,$$

  where $\mathbf{Set}_{\mathrm{B-E}}$ is the set of assumptions listed in Table 3.1, $\mathcal{A}$ is the following model formulation

$$\rho \mathbf{F} \frac{\partial^2 u}{\partial t^2} + E I_y \frac{\partial^4 u}{\partial x^4} = 0,$$

  which is obtained under the formalisation mapping S_1 described during construction of the category **Beam**.
- A mathematical model of the Rayleigh beam theory is the triple

$$\mathfrak{M}_{\mathrm{R}} := \langle \mathbf{Set}_{\mathrm{R}}, \mathcal{B}, S_1 \rangle,$$

where $\mathbf{Set}_\mathrm{R}$ is the set of assumptions listed in Table 3.1, $\mathcal{B}$ is the following model formulation

$$\rho F \frac{\partial^2 u}{\partial t^2} + E I_y \frac{\partial^4 u}{\partial x^4} - \rho I_y \frac{\partial^4 u}{\partial x^2 \partial t^2} = 0,$$

which is obtained under the formalisation mapping S_1 described during construction of the category **Beam**.

- A mathematical model of the Timoshenko beam theory is the triple

$$\mathfrak{M}_\mathrm{T} := \langle \mathbf{Set}_\mathrm{T}, \mathcal{C}, S_1 \rangle,$$

where $\mathbf{Set}_\mathrm{T}$ is the set of assumptions listed in Table 3.1, $\mathcal{C}$ is the following model formulation

$$\begin{cases} \rho F \dfrac{\partial^2 u}{\partial t^2} - \aleph \mu F \dfrac{\partial^2 u}{\partial x^2} + \aleph \mu F \dfrac{\partial \varphi}{\partial x} & = 0, \\ \rho I_y \dfrac{\partial^2 \varphi}{\partial t^2} - E I_y \dfrac{\partial^2 \varphi}{\partial x^2} + \aleph \mu F \left(\varphi - \dfrac{\partial u}{\partial x} \right) & = 0, \end{cases}$$

which is obtained under the formalisation mapping S_1 described during construction of the category **Beam**.

While discussing the category **Beam**, it has been pointed out that there are two possible differential formulations of the Timoshenko beam theory. Within the category theory-based modelling framework, this situation has been resolved through a natural transformation of the formalisation mapping S. In this more general approach, however, we need to introduce another model for this situation: Another mathematical model of Timoshenko beam theory is the triple

$$\mathfrak{M}_\mathrm{T}^{'} := \langle \mathbf{Set}_\mathrm{T}, \mathcal{D}, S_2 \rangle,$$

where $\mathbf{Set}_\mathrm{T}$ is the set of assumptions listed in Table 3.1, $\mathcal{D}$ is the following model formulation

$$\rho F \frac{\partial^2 u}{\partial t^2} + E I_y \frac{\partial^4 u}{\partial x^4} - \rho I_y \left(1 + \frac{E}{\aleph \mu} \right) \frac{\partial^4 u}{\partial x^2 \partial t^2} + \frac{\rho^2 I_y}{\aleph \mu} \frac{\partial^4 u}{\partial t^4} = 0,$$

which is obtained under the formalisation mapping S_2. In fact, the only difference of the formalisation mapping S_2 to S_1 is that, after obtaining a model formulation $\mathcal{C}$, the system of equations is reduced to a single differential equation. Nonetheless, it is necessary to address these two models separately within this general framework.

Another difference to the category theory-based approach is that there is no *mathematical structure* "covering" the models $\mathfrak{M}_{\mathrm{B-E}}$, $\mathfrak{M}_\mathrm{R}$, $\mathfrak{M}_\mathrm{T}$, and $\mathfrak{M}_\mathrm{T}^{'}$. In other words, these models are not related by the construction. On one hand, this has an obvious disadvantage of losing a formal structure, but on the other hand, it provides new possibilities for introducing model relations, and, consequently, makes the approach based on Definition 5.3 more flexible.

To compensate for missing structure, the following two alternatives for working with mathematical models could be used:

1. The concept of a lattice of theories, see again [28], can be adapted, as it has been done already for coupling of models from multiple categories in Sect. 3.3.5.
2. The abstract relational algebra approach originated from works [15, 17, 22, 29] can be adapted.

Let us further explore both alternatives, and we will see that they are not truly alternatives but rather two branches of one approach to introducing a formal structure for mathematical models in the sense of Definition 5.3.

5.4.2 *Lattice of Mathematical Models*

Let us now illustrate the adaptation of the lattice of theories concept to mathematical models using the example of four models associated with the classical beam theories: $\mathfrak{M}_{\text{B-E}}$, $\mathfrak{M}_{\text{R}}$, $\mathfrak{M}_{\text{T}}$, and $\mathfrak{M}'_{\text{T}}$. Since no formal relations between these models have been introduced yet, there are several possibilities for creating a lattice of these models. Figure 5.1 presents the lattice of mathematical models with expansions and analogy operations being used on the lattice. As it can be clearly observed, this figure represents a kind of blending between the diagrams depicting the structures of the categories **Beam** and **Beam**^{F}. Furthermore, in constructing this lattice of mathematical models, we have *implicitly* adapted the idea of relating models by comparing their sets of assumptions, as described through the concept of complexity in Definition 3.2. In this context, the expansion operation is understood as expanding the range of physical effects and phenomena describable by a model, while the analogy operation is understood as a reformulation of the model formulation C into $\mathcal{D}$.

$$\begin{array}{ccccc} & & \mathfrak{M}_{\text{R}} & & \\ & \overset{\mathbf{E}}{\nearrow} & & \overset{\mathbf{E}}{\searrow} & \\ \mathfrak{M}_{\text{B-E}} & & & & \mathfrak{M}_{\text{T}} \xrightarrow{\mathbf{A}} \mathfrak{M}'_{\text{T}} \end{array}$$

Fig. 5.1 An example of a lattice of mathematical models with expansion and analogy operations. This lattice implicitly uses the complexity relation introduced in Definition 3.2

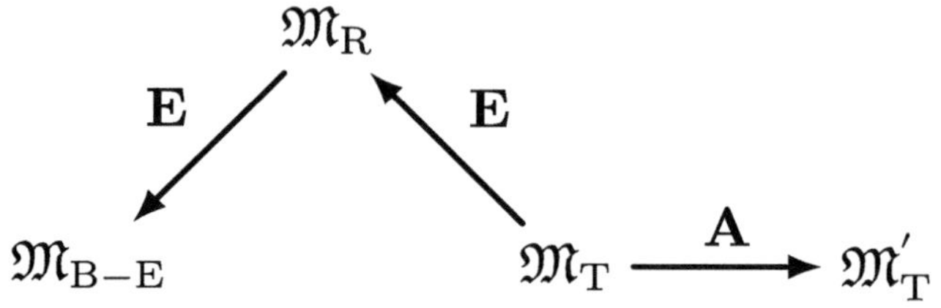

Fig. 5.2 Another example of a lattice of mathematical models with expansion and analogy operations. In this case, expansion is understood as a "simple" expansion of sets, independent of any modelling context

Another possibility for creating a lattice of mathematical models is shown in Fig. 5.2, where expansion and analogy operations have also been used. However, in this case, all the expansion arrows are reverse. Hence, the expansion operation is understood as a direct expansion of the sets of assumptions. This means that the sets $\mathbf{Set}_{\mathrm{B-E}}$, $\mathbf{Set}_{\mathrm{R}}$, and $\mathbf{Set}_{\mathrm{T}}$ are compared and if $\mathbf{Set}_{\mathrm{T}} \subset \mathbf{Set}_{\mathrm{R}}$ holds, it implies that $\mathbf{Set}_{\mathrm{T}}$ has been expanded. As we see, this interpretation is entirely different from the one based on the complexity definition and, in fact, is even free of any modelling context. However, the analogy operation is still interpreted as described above.

The examples of lattices of mathematical models shown in Figs. 5.1 and 5.2 demonstrate the flexibility of this approach. At the same time, these examples reveal that the interpretation of these lattices is purely subjective, as is typically the case for any knowledge representation system. However, it is important to remark, that this subjectivity can be eliminated if all movements along the lattice are clearly defined, as in the classical case of the lattice of theories from [28]. Keeping this idea in mind, let us now examine another example: we will create a lattice of mathematical models associated with the objects of the category **Elast**.

Let us now define various models of linear elasticity associated with the objects of **Elast** as triples:

- A mathematical model of the linear elasticity is the triple

$$\mathfrak{M}_{\mathrm{LE}} := \langle \mathbf{Set}_{\mathrm{LE}}, \mathcal{A}, S_1 \rangle,$$

where $\mathbf{Set}_{\mathrm{LE}}$ is the set of assumptions listed in Table 3.2, $\mathcal{A}$ is the following model formulation

$$\mathbf{Set}_{\mathrm{LE}} \quad \overset{S}{\mapsto} \quad \begin{cases} \sigma_{ij} = \lambda \delta_{ij} \varepsilon_{kk} + 2\mu \varepsilon_{ij} \\ \varepsilon = \dfrac{1}{2}(\nabla \mathbf{u} + \nabla \mathbf{u}^T) \\ \nabla \cdot \boldsymbol{\sigma} + \rho \mathbf{F} = 0 \end{cases} \quad =: \mathcal{A}.$$

which is obtained under the formalisation mapping S_1 described during construction of the category **Elast**.

- A mathematical model of the micropolar elasticity is the triple

$$\mathfrak{M}_{\text{ME}} := \langle \mathbf{Set}_{\text{ME}}, \mathcal{B}, S_1 \rangle,$$

where $\mathbf{Set}_{\text{ME}}$ is the set of assumptions listed in Table 3.2, $\mathcal{B}$ is the following model formulation

$$\mathbf{Set}_{\text{ME}} \overset{S}{\mapsto} \begin{cases} \sigma_{ji} = (\mu + \alpha)\varepsilon_{ji} + (\mu - \alpha)\varepsilon_{ij} + \lambda\delta_{ij}\varepsilon_{kk} \\ \mu_{ji} = (\gamma + \epsilon)\kappa_{ji} + (\gamma - \epsilon)\kappa_{ij} + \beta\delta_{ij}\kappa_{kk} \\ \varepsilon_{ji} = u_{i,j} - e_{kij}\varphi_k, \quad \kappa_{ji} = \varphi_{i,j}, \quad \boldsymbol{\varphi} = \frac{1}{2}\nabla \times \mathbf{u} \\ \sigma_{ji,j} + f_i = 0 \\ e_{ijk}\sigma_{jk} + \mu_{ji,j} + c_i = 0 \end{cases} =: \mathcal{B}.$$

which is obtained under the formalisation mapping S_1 described during construction of the category **Elast**.
- A mathematical model of the nonlinear elasticity is the triple

$$\mathfrak{M}_{\text{NE}} := \langle \mathbf{Set}_{\text{NE}}, C, S_1 \rangle,$$

where $\mathbf{Set}_{\text{NE}}$ is the set of assumptions listed in Table 3.2, C is the following model formulation

$$\mathbf{Set}_{\text{NE}} \overset{S}{\mapsto} \begin{cases} \boldsymbol{\sigma} = \lambda\mathbf{I} + 2\mu\mathbf{C} \\ \mathbf{C} = \mathbf{F}^T\mathbf{F} \\ \nabla \cdot \boldsymbol{\sigma} + \rho\mathbf{K} = 0 \end{cases} =: C.$$

which is obtained under the formalisation mapping S_1 described during construction of the category **Elast**.
- A mathematical model of the linear elastic fracture mechanics is the triple

$$\mathfrak{M}_{\text{LEF}} := \langle \mathbf{Set}_{\text{LEF}}, \mathcal{D}, S_1 \rangle,$$

where $\mathbf{Set}_{\text{LEF}}$ is the set of assumptions listed in Table 3.2, $\mathcal{D}$ is the following model formulation

$$\mathbf{Set}_{\text{LEF}} \overset{S}{\mapsto} \begin{cases} \sigma_{ij} = \lambda\delta_{ij}\varepsilon_{kk} + 2\mu\varepsilon_{ij} \\ \varepsilon = \frac{1}{2}(\nabla\mathbf{u} + \nabla\mathbf{u}^T) \\ \nabla \cdot \boldsymbol{\sigma} + \rho\mathbf{F} = 0 \\ \text{Cracks geometric information} \end{cases} =: \mathcal{D}.$$

which is obtained under the formalisation mapping S_1 described during construction of the category **Elast**.

Let us now discuss the creation of a lattice of models $\mathfrak{M}_{\mathrm{LE}}$, $\mathfrak{M}_{\mathrm{ME}}$, $\mathfrak{M}_{\mathrm{NE}}$, and $\mathfrak{M}_{\mathrm{LEF}}$. This example will illustrate why the approach based on Definition 5.3 is more flexible, but less rigorous. If we implicitly follow the idea of comparing models based on their complexity, then the lattice of mathematical models presented in Fig. 5.3 will be obtained. Evidently, this lattice has the same structure as the diagram of the category **Elast**.

However, perhaps the lattice shown in Fig. 5.4 better describes the structure between these four theories. In general, it is also possible to use expansion operation, for example, between $\mathfrak{M}_{\mathrm{ME}}$ and $\mathfrak{M}_{\mathrm{NE}}$. However, the interpretation of these expansions is rather not straightforward, because we need at first to discard some axioms from the micropolar theory and then add new axioms from the non-linear elasticity. Hence, we use the revision operation, as depicted in Fig. 5.4. Evidently, this interpretation is closely aligned with the original concept of the lattice of theories.

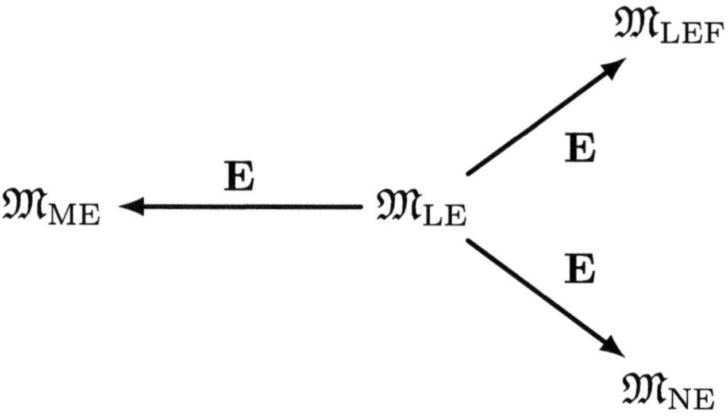

Fig. 5.3 Lattice of elasticity models, where the expansion operations are defined based on the complexity relation introduced in Definition 3.2

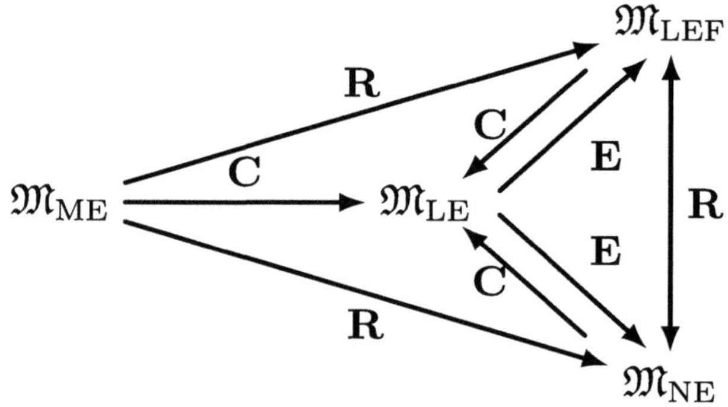

Fig. 5.4 Lattice of elasticity theories, where movements along the lattice are interpreted closer to the original concept of the lattice of theories

Let us now summarise the key points of the concept of lattices of mathematical models:

- To construct a lattice of mathematical models, it is necessary to have a clear idea of how movements along the lattice are to be interpreted;
- Formally, it is always possible to draw an "expansion arrow" from any model to any other model. But this approach is impractical, because it "destroys" the clarity of the lattice.
- While a clear interpretation of the movements along the lattice is essential, the choice of these interpretations is still free, which enhances the flexibility of this approach.

As it has been demonstrated by the example presented above, it is essential to have a clear understanding of how the movements along the lattice should be interpreted. In the next section, we will discuss a relational algebra approach, which, in fact, complements the concept of lattices of mathematical models.

5.4.3 Relational Algebra Approach

The results on the concept of lattice of mathematical models illustrate that by waiving the categorical structure for describing mathematical models and the objects related to them, we gain more flexibility. However, at the same time, a clear structure of relations between the models is lost. To compensate for this missing structure, we propose enriching the framework of Definition 5.3 by including a relational algebra approach. This approach is motived by works [15, 17, 22, 29]. In this context, we introduce the following definition on top of Definition 5.3:

Definition 5.4 The object $\mathfrak{M}_{\mathfrak{C}} = \langle \mathfrak{M}^N, \mathfrak{R} \rangle$ is called a *collection of mathematical models*, where

- $\mathfrak{M}^N = (\mathfrak{M}_1, \mathfrak{M}_2, \ldots, \mathfrak{M}_N)$ is an N-tuple of mathematical models in the sense of Definition 5.3;
- $\mathfrak{R} = (R_1, R_2, \ldots, R_m)$ is an m-tuple of relations.

With this definition, instead of imposing a strict structure in terms of categories, we use the tuple of relations $\mathfrak{R}$ to structure different mathematical models. Evidently, this approach provides more flexibility for relating models, as multiple binary relations can be considered simultaneously, if necessary. Additionally, in contrast to the category theory-based approach, models describing different physical phenomena can be put together into a single collection of mathematical models. Furthermore, a more formal structure, in comparison to the concept of lattice of mathematical models, has been introduced by the tuple of relations $\mathfrak{R}$.

To illustrate how the approach based on Definition 5.4 can be used in practice, let us consider the following collection of mathematical models:

$$\mathfrak{M}_{\mathfrak{C}} = \left\langle \left(\mathfrak{M}_{\mathrm{B-E}}, \mathfrak{M}_{\mathrm{R}}, \mathfrak{M}_{\mathrm{T}}, \mathfrak{M}_{\mathrm{T}}^{'}, \mathfrak{M}_{\mathrm{LE}}, \mathfrak{M}_{\mathrm{ME}}, \mathfrak{M}_{\mathrm{NE}}, \mathfrak{M}_{\mathrm{LEF}} \right), (R_1, R_2) \right\rangle,$$

where we have listed all the models discussed in the previous subsection. The goal is now to establish a lattice of mathematical models, where the expansion operation implies satisfaction of at least one of the relations from the couple (R_1, R_2), where

- R_1 is understood in the sense of complexity introduced in Definition 3.2;
- R_2 implies that one model is a one-dimensional special case of another models.

For indicating the basis of an expansion operation, we will use subscripts: $\mathbf{E}_{R_1}$ and $\mathbf{E}_{R_2}$. Thus, the lattice of mathematical models shown in Fig. 5.5 is obtained.

As we can see from Fig. 5.5, the relation R_2 helps establish a connection between the one-dimensional beam theories and elasticity theories. In a broader context, considering different binary relations in an m-tuple $\mathfrak{R}$ allows us to relate models that do not necessarily describe the same physical phenomenon and are not necessary being coupled together, as it is required in the category theory-based modelling framework. Therefore, in some sense, the ideas described in this section are generalisations of the category theory-based modelling framework from Chap. 3. Nonetheless, it is important to emphasise that both approaches to the formalisation of mathematical modelling complement each other, and the choice of a specific

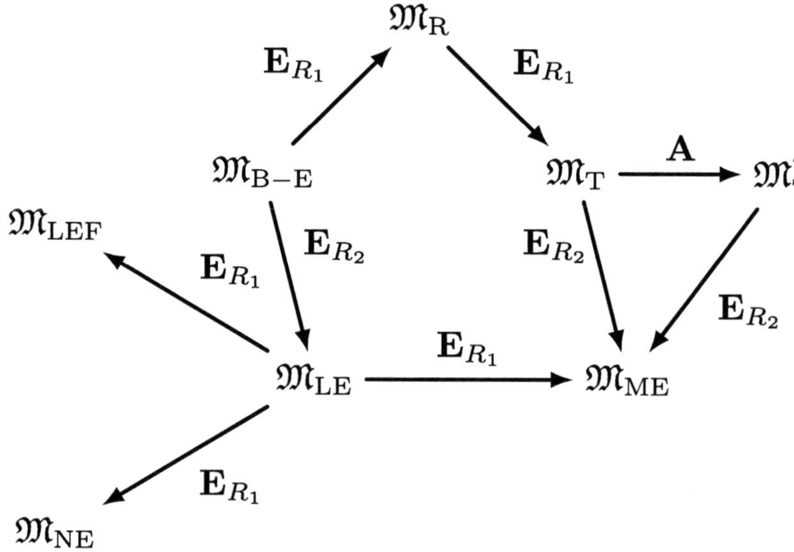

Fig. 5.5 A lattice of mathematical models that belong to a single collection and are subject to two relations

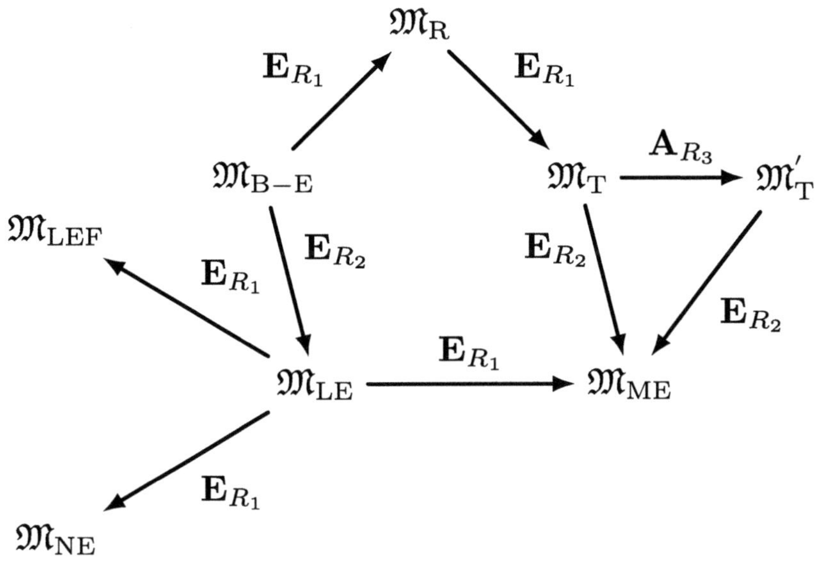

Fig. 5.6 Lattice of mathematical models, which belong to a single collection and are subject to three relations: relations R_1 and R_2 are associated with expansions, while relation R_3 is associated with analogy

approach for practical tasks depends on the concrete problem formulations and current objectives.

Additionally, relations can also be associated with analogy operations on a lattice of mathematical models. For example, if we extend the pair of relations (R_1, R_2) by the relation R_3, which holds if and only if two model formulations can be converted into one another, then the analogy shown in Fig. 5.5 will be also sub-indexed with R_3. Figure 5.6 illustrates this modification of the lattice of mathematical models.

The next step is to extended the ideas presented above to the case of coupling of mathematical models. The idea here is to follow the same strategy discussed around Figs. 5.5 and 5.6 and extend the tuple of relations for addressing the coupling of models. It is important to emphasise again that, in contrast to the category theory-based approach, we do not have a general "umbrella", i.e. categories, that subdivides models into coupled models and those being coupled. Therefore, we need to put all models, including coupled models, in one collection according to Definition 5.4.

For illustrative purposes, we will now consider three models of thermoelasticity, which have been discussed in the context of category **ThermoElast**, see Sect. 3.3.4.2. Since there are in total 36 different coupled models, we focus on one special case. The goal here is demonstrate how the coupling of models can be represented in the current setting and highlight the differences compared to the category theory-based approach. Therefore, we will consider the following collection of mathematical models:

$$\mathfrak{M}_{\mathfrak{C}}^{\text{TherEl}} = \left\langle \left(\mathfrak{M}_{\text{UN}}, \mathfrak{M}_{\text{LE}}, \mathfrak{M}_{\text{LE,UN}}^{\text{CDT}}, \mathfrak{M}_{\text{LE,UN}}^{\text{CQST}}, \mathfrak{M}_{\text{LE,UN}}^{\text{DT}} \right), (R_1, R_2) \right\rangle,$$

where $\mathfrak{M}_{\text{UN}}$ is the model of heat conduction in the case of a uniform material, $\mathfrak{M}_{\text{LE}}$ is the model of linear elasticity, $\mathfrak{M}_{\text{LE,UN}}^{\text{CDT}}$ is the model of coupled dynamic thermoelasticity, $\mathfrak{M}_{\text{LE,UN}}^{\text{CQST}}$ is the model of coupled quasi-static thermoelasticity, $\mathfrak{M}_{\text{LE,UN}}^{\text{DT}}$ is the model of dynamic thermoelasticity, see the discussions around categories **Heat**, **Elast**, and **ThermoElast** for details about the corresponding sets of assumptions and model formulations. For the couple of relations (R_1, R_2), we consider the following concrete cases:

- R_1 is true, when a model is part of a coupled model;
- R_2 is understood in the sense of coupling complexity introduced in Definition 3.8.

Figure 5.7 presents the lattice of mathematical models corresponding to the collection $\mathfrak{M}_{\mathfrak{C}}^{\text{TherEl}}$.

This figure can be extended to the case of more models, but of course, the readability of the lattice will decrease. To enhance the clarity of larger lattices, multiple mathematical models might be grouped into different collections, with the relations between these collections shown in a diagram. However, in this case, we are more or less mimicking the ideas of category theory-based approach without explicitly naming it. Nonetheless, because relations can be arbitrary, the relational algebra approach still provides greater flexibility.

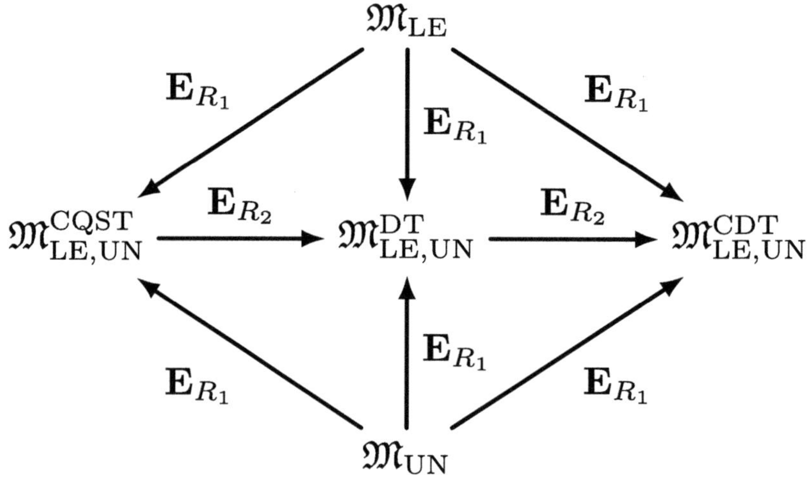

Fig. 5.7 Lattice of mathematical models in the case of coupling models describing different physical phenomena

The case of coupling mathematical models that describe the same physical phenomenon is more interesting to discuss in the current setting. We will again use the example of coupling heat conduction models for a domain made of two different materials:

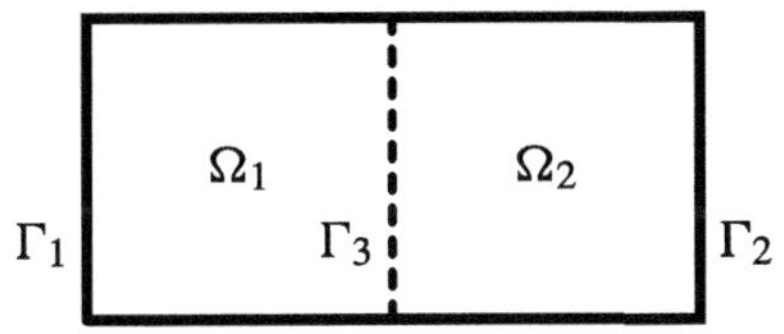

For illustrative purposes, let us consider only two heat conduction models: the model of heat conduction in the case of a uniform material, denoted by $\mathfrak{M}_{\text{UN}}$, and the model of heat conduction in the case of a radiative heat exchange with the surroundings, denoted by $\mathfrak{M}_{\text{RH}}$. For details on the models and their assumptions, we refer to the discussion on the category **HeatCoup** in Sect. 3.3.4.1. To represent the coupling of these models with each other, the following collection of mathematical models needs to be considered:

$$\mathfrak{M}_{\mathfrak{C}}^{\text{Heat}} = \left\langle \left(\mathfrak{M}_{\text{UN}}, \mathfrak{M}_{\text{RH}}, \mathfrak{M}_{\text{UN,UN}}, \mathfrak{M}_{\text{UN,RH}}, \mathfrak{M}_{\text{RH,UN}}, \mathfrak{M}_{\text{RH,RH}}\right), (R_1, R_2)\right\rangle,$$

where the relation R_1 is true then a model is part of a coupled model in subdomain Ω_1, and the relation R_2 is true when a model is part of a coupled model in subdomain Ω_2. Note that these relations reflect the idea of the co-product structure, which has been introduced for coupling of mathematical models from the same category in Definition 3.7.

Figure 5.8 presents the lattice of mathematical models corresponding to the collection $\mathfrak{M}_{\mathfrak{C}}^{\text{Heat}}$. Here, two extensions $\mathbf{E}_{R_1}$ and $\mathbf{E}_{R_2}$, indicate that the same mathematical model has been used for the coupling: one for Ω_1 and one for Ω_2.

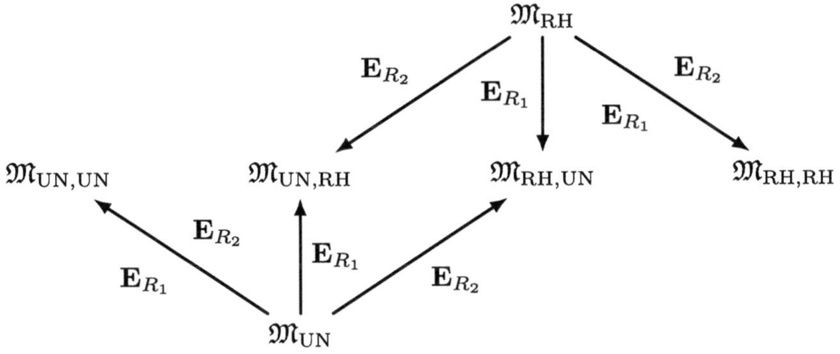

Fig. 5.8 Lattice of mathematical models for the coupling of models describing the same physical phenomenon

Similar to the case of category theory-based approach, it is formally necessary to distinguish between models $\mathfrak{M}_{\mathrm{UN,RH}}$ and $\mathfrak{M}_{\mathrm{RH,UN}}$, even though the same models have been coupled, since their acting subdomains are different.

Again, we see that the approach based on relational algebra combined with the concept of lattice of mathematical models provides a high level of flexibility, as relations can be chosen arbitrarily. However, this flexibility comes at the cost of loosing the strict structure found in the category theory-based approach. Consequently, it is not correct to claim that one approach is better than another; rather, it is beneficial to consider both as complementary ideas for better understanding and representing the modelling process.

It is also interesting to make the following remark:

Remark 5.6 The concept of lattice of mathematical models, combined with the relational algebra approach, can be further enriched by interpreting these constructions from the point of view of the category **Grph** of all small graphs, see [20] for details.

5.4.4 Extension of the Relational Algebra Approach Beyond Mathematical Models

In the final part of this section, we illustrate that the relational algebra approach can be generalise to allow the conceptual describing of not only mathematical models but also general engineering systems. For instance, in [15, 17] this approach has been applied to formalise intelligent structures and autonomous construction sites, respectively. It is important to note that the application of the relational algebra approach to the conceptual modelling of engineering systems has been inspired by works [22, 27, 29].

The overall idea of using the relational algebra approach to describe engineering systems is to capture the only essential parts of it while leaving out unnecessary details at first. Once the essence of the system is captured, it can be further enriched with additional details. Following this philosophy, first ideas of using abstract algebraic definitions to describe engineering systems have been presented in [17]. In this work, intelligent structures, composed of a *physical domain* and a *cyber domain*, see for example [13] and references therein, have been abstractly described.

Without delving into the details of the specific engineering problem addressed in [17], let us recall a few abstract definitions to illustrate how the relational algebra approach can support conceptual modelling in practice. Sensors, sensor clusters, and sensor networks are essential part of an intelligent structure, or, more generally, any structure equipped with a structural health monitoring system. Accordingly, we introduce the following abstract definitions of these objects:

Definition 5.5 A sensor is the object $\mathcal{S} = \langle \mathcal{I}, \mathcal{Y}, \mathcal{T} \rangle$, where

- $\mathcal{I} = (I_1, I_2, \ldots, I_n)$ is an n-tuple of finite index sets;
- $\mathcal{Y} = (Y_1, Y_2, \ldots, Y_n)$ is an n-tuple of measurements with $Y_i \in \mathbb{R}^{N_i}$, $i = 1, \ldots, n$;
- $\mathcal{T}$ is an k-tuple of specifications (typing information).

Definition 5.6 A sensor cluster is the object $\mathcal{S}_C = \langle \mathfrak{B}, \mathfrak{S}, \mathfrak{R} \rangle$, where

- $\mathfrak{B}$ is a sensor node or a base station controlling the sensor cluster;
- $\mathfrak{S} = (\mathcal{S}_1, \mathcal{S}_2, \ldots, \mathcal{S}_n)$ is an n-tuple of sensors, introduced in Definition 5.5;
- $\mathfrak{R} = (R_1, R_2, \ldots, R_m)$ is an m-tuple of relations.

Moreover, a sensor cluster can be represented by help of the following composition:

$$\mathcal{S}_C := \mathcal{S}_1 \underset{R_1}{\circ} \mathcal{S}_2 \underset{R_2}{\circ} \cdots \underset{R_m}{\circ} \mathcal{S}_n, \tag{5.14}$$

where the relations $R_1, R_2, \ldots, R_m$ are defined by the base station $\mathfrak{B}$.

Definition 5.7 A sensor network is the object $\mathcal{S}_\mathcal{N} = \langle \mathfrak{S}_\mathfrak{N}, \mathfrak{R} \rangle$, where

- $\mathfrak{S}_\mathfrak{N} = \left(\mathcal{S}_C^1, \mathcal{S}_C^2, \ldots, \mathcal{S}_C^n\right)$ is an n-tuple of sensor clusters, introduced in Definition 5.6;
- $\mathfrak{R} = (R_1, R_2, \ldots, R_k)$ is an k-tuple of relations.

Thus, a sensor network can be represented by help of the following composition:

$$\mathcal{S}_\mathcal{N} := \mathcal{S}_C^1 \underset{R_1}{\circ} \mathcal{S}_C^2 \underset{R_2}{\circ} \cdots \underset{R_k}{\circ} \mathcal{S}_C^n,$$

where the relations $R_1, R_2, \ldots, R_k$ specify communication among different bases stations of sensor clusters $\mathcal{S}_C^i$, $i = 1, 2, \ldots, n$.

The idea of Definition 5.6 is to underline that sensors combined into a sensor cluster communicate with the base station, as well as sensors can (optionally) communicate among themselves. The communication rules are set by the relations $R_1, R_2, \ldots, R_m$, which are specified during the sensor network design, see [22, 29] for specific examples of relations and practical meanings of the relations in wireless sensor network modelling. Thus, these relations serve as composition rules in (5.14). Moreover, considering that communication rules can be set individually for each sensor or generally for the whole system, the number of relations m is not related to the number of sensors n constituting the sensor cluster.

Similar to Definitions 5.5–5.7, actuators and actuator networks were defined in [17]. One of the main motivations for these definitions at the time was to promote a compositional perspective on engineering systems. Consequently, an intelligent structure was defined as a composition of sensor and actuator networks. Later, in [15], it was recognised that combining abstract algebraic definitions with categorical ontology logs, see Sect. 2.4, provides greater expressive power. Consequently, this combination was applied to the conceptual description of autonomous construction sites in [15]. Moreover, the lattice of theories concept was used as well, and its advantages were discussed from an engineering perspective.

Let us elaborate on the ideas of combining abstract definitions with ologs by recalled the definition of an autonomous construction site from [15]:

Definition 5.8 (Autonomous Construction Site) An *autonomous construction site* or a *robotic environment* is the object $\mathfrak{A} = \langle \mathfrak{T}, \mathfrak{R}, \mathfrak{H}, \mathfrak{E}, \mathfrak{G}, \mathfrak{B} \rangle)$, where

- $\mathfrak{T}$ is a task to be solved by robots on a construction site;
- $\mathfrak{R} = (R_1, R_2, \ldots, n)$ is an n-tuple of robots;
- $\mathfrak{H}$ is the object describing human-robot interaction on a construction site;
- $\mathfrak{E}$ is a set of pairs representing environmental conditions on a construction site;
- $\mathfrak{G}$ is a 4-tuple of a GPS information for important parts of a construction site;
- $\mathfrak{B}$ is a base station controlling the autonomous construction site.

Figure 5.9 present the olog representation of this definition. Note that, this olog can be expanded with more details, if required, by using the expansion operation to move along the lattice of representations.

Thus, we see that the relational algebra approach can be applied to the conceptual modelling of various engineering systems, which are not necessarily described immediately in terms of mathematical expressions and equations. This point has

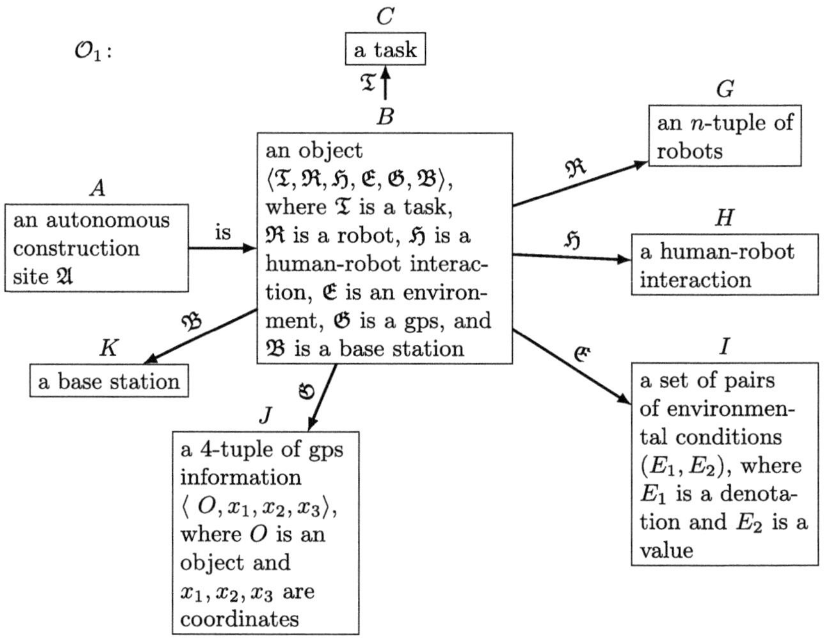

Fig. 5.9 Olog representation of an autonomous construction site introduced in Definition 5.8, see [15] for details

led to the introduction of a *meta-metamodelling approach*,—combining ologs, the lattice of theories concept, and the relational algebra approach. The term "meta-metamodelling" has been chosen because "metamodelling" typically refers to the use of tools such as the Unified Modelling Language (UML) or other so called conceptual modelling languages [3]. However, the ideas presented in this section are more general, allowing for very abstract descriptions of objects. Consequently, these ideas address the meta-meta level.

5.5 Summary of the Chapter

In this chapter, we have further extended the ideas of the category theory-based modelling framework introduced in Chap. 3. Specifically, we have revisited the definition of categories of mathematical models and presented initial concepts aimed at automatic model generation. Along this path, the notion of a consistently formalisable set of assumptions has been proposed. The critical point here is the theoretical existence of a verification algorithm allowing checking the derivation of a model from its underlying set of assumptions. This problem can be seen as analogous to the decidability problem in logic or the decision problem in computer science. Thus, the ideas on automatic model generation open potential directions for future research at the intersection of mathematics and theoretical computer science.

In Sect. 5.2, we have explored the ideas of using type theory to formalise mathematical models and to develop an algorithm for verifying the consistency of model derivation. In this case, a particular types are assigned to each basic model component, and a complete model is constructed by composing these types. The advantage of using type theory is that many programming languages utilise the idea of composing types. As a result, the type checker of a programming language can be used to verify the model derivation.

The use of type theory for conceptual modelling is based on the FHM (Functional Hybrid Modelling). Originally, FHM has been developed to address mathematical models formulated as differential-algebraic equations. Evidently, problems in mathematical physics and engineering often require PDEs. To address this need, an extension of FHM to PDEs has been proposed in Sect. 5.2. This section concludes with the presentation of a general algorithm for verifying model derivations. It is important to note that these ideas are still in their early stages and provide preliminary insights for future research and discussion.

The discussion on using types to formalise mathematical models has naturally led to a categorical description of consistency of coupling, which has been presented in Sect. 5.3. This description addresses the categories of formalised modelling assumptions, i.e. categories $\mathbf{Model}^F$.

Finally, an alternative perspective on mathematical models as abstract triples has been presented Sect. 5.4. This perspective captures the essence of the categorical description of mathematical models while removing all structural constrains. To compensate for the missing structure, two approaches have been discussed:

1. the concept of a lattice of mathematical models; and
2. the relational algebra approach.

Both approaches have been discussed and illustrated with examples, leading to the observation that their combination provides greater flexibility in expressing mathematical models and the relations between them. Furthermore, the integration of the relational algebra approach with ologs and the lattice of theories concept enables the description of arbitrary engineering systems, which are not necessarily immediately expressed in terms of mathematical expressions and equations.

References

1. Barthe, G., & Coquand, T. (2002). An introduction to dependent type theory. In G. Barthe, P. Dybjer, L. Pinto, & J. Saraiva (Eds.), *Applied Semantics. APPSEM 2000*. Lecture notes in computer science (Vol. 2395). Springer.
2. Bird, R. (2015). *Thinking functionally with Haskell*. Cambridge University Press.
3. Borgida, A. (1986). Conceptual modeling of information systems. In M. L. Brodie, J. Mylopoulos, & J. W. Schmidt (Eds.), *Topics in information systems – On knowledge base management systems* (pp. 461–469). Springer.
4. Capper, J. (2014). *Semantic methods for functional hybrid modelling*. PhD thesis, University of Nottingham.
5. Capper, J., & Nilsson, H. (2011). Static balance checking for first-class modular systems of equations. In R. Page, Z. Horváth, & V. Zsók (Eds.), *Trends in Functional Programming: 11th International Symposium, TFP 2010*, Norman, May 2010. Revised Selected Papers. Lecture notes in computer science (Vol. 6546, pp. 50–65). Springer.
6. Capper, J., & Nilsson, H. (2012). Structural types for systems of equations: Type refinements for structurally dynamic first-class modular systems of equations. *Higher-Order and Symbolic Computation, 25*(2–4), 275–310.
7. Chaudhry, H. (2021). *Enriching a type-theoretic modelling methodology by dimensional types*. Master Thesis, Bauhaus-Universität Weimar.
8. Gürlebeck, K., & Sprößig, W. (1997). *Quaternionic and Clifford calculus for physicists and engineers*. John Wiley & Sons.
9. Gürlebeck, K., Hofmann, D., & Legatiuk, D. (2017). Categorical approach to modelling and to coupling of models. *Mathematical Methods in the Applied Sciences, 40*(3), 523–534.
10. Gürlebeck, K., Legatiuk, D., Nilsson, H., & Smarsly, K. (2019). Conceptual modelling: Towards detecting modelling errors in engineering applications. *Mathematical Methods in the Applied Sciences, 43*(3), 1243–1252.
11. Hughes, G. E., & Cresswell, M. J. (1984). *A companion to modal logic*. Methuen & Co. Ltd.
12. Jhala, R., Seidel, E., & Vazou, N. (2017). *Programming with refinement types. An introduction to LiquidHaskell*. https://ucsd-progsys.github.io/liquidhaskell-blog/
13. Korkmaz, S. (2011). A review of active structural control: Challenges for engineering informatics. *Computers and Structures, 89*(23), 2113–2132.
14. Kroening, D., & Strichman, O. (2008). *Decision procedures. An algorithmic point of view*. Springer.
15. Legatiuk, D., & Luckey, D. (2023). Formalising autonomous construction sites by help of abstract mathematics. *Eng, 4*(1), 799–815.
16. Legatiuk, D., & Nilsson, H. (2017). Abstract modelling: Towards a typed declarative language for the conceptual modelling phase. In *Proceedings of 8th International Workshop on Equation-Based Object-Oriented Modeling Languages and Tools*, Weßling.

17. Legatiuk, D., & Smarsly, K. (2018). An abstract approach towards modeling intelligent structural systems. In *Proceedings of the 9th European Workshop on Structural Health Monitoring*, Manchester.
18. Loveland, D. W., Hodel, R. E., & Sterrett, S. G. (2014). *Three views of logic. Mathematics, philosophy, and computer science*. Princeton University Press.
19. Lurie, A. I. (2005). *Theory of elasticity*. Foundations of engineering mechanics. Springer.
20. Mac Lane, S. (1978). *Categories for the working mathematician*. Springer.
21. Muskhelishvili, N. I. (1977). *Some basic problems of the mathematical theory of elasticity*. Springer Science+Business Media Dordrecht.
22. Nefzi, B., Schott, R., Song, Y. Q., Staples, G. S., & Tsiontsiou, E. (2015). An operator calculus approach for multi-constrained routing in wireless sensor networks. In *Proceedings of the 16th ACM International Symposium on Mobile Ad Hoc Networking and Computing*, New York, June 22–25.
23. Nilsson, H., Curtney, A., & Peterson, J. (2002). Functional reactive programming, continued. In *Proceedings of the 2002 ACM SIGPLAN Haskell Workshop* (pp. 51–64).
24. Nilsson, H., Peterson, J., & Hudak, P. (2003). Functional hybrid modeling. In *Lecture notes in computer science* (Vol. 2562, pp. 376–390). Springer.
25. Norell, U., & Chapman, J. (2008). Dependently typed programming in Agda. In *Proceeding AFP'08 Proceedings of the 6th International Conference on Advanced Functional Programming*, Heijen, May 19–24 (pp. 230–226).
26. Pierce, B. C. (2002). *Types and programming languages*. MIT Press.
27. Simko, G., Levendovszky, T., Maroti, M., & Sztipanovits, J. (2014). Towards a theory for cyber-physical systems modeling. In *Proceedings of the 4th ACM SIGBED International Workshop on Design, Modeling, and Evaluation of Cyber-Physical Systems*, Berlin, April 14.
28. Sowa, J. (2000). *Knowledge representation: Logical, philosophical, and computational foundations*. Brooks/Cole.
29. Syarif, A., Abouaissa, A., Idoumghar, L., Lorenz, P., Schott, R., & Staples, S. G. (2019). New path centrality based on operator calculus approach for wireless sensor network deployment. *IEEE Transactions on Emerging Topics in Computing, 7*(1), 162–173.
30. Vassilyev, S. N., Druzhinin, A. E., & Morozov, N. Yu. (2015). Derivation of preservation conditions for properties of mathematical models. *Doklady Mathematics, 92*(3), 658–663.

Chapter 6
Epilogue

The growing complexity of engineering problems necessitates the development of advanced modelling approaches that address not only the correctness of numerical treatment of models but also the process of model creation. In particular, mathematical models are based on modelling assumptions, which can be violated even before computing with the model. Thus, it is necessary to distinguish between conceptual modelling errors, related to breaking basic modelling assumptions, and modelling errors related to practical realisation of models, such as the quality of computer implementation and approximation quality. While the errors related to practical realisation of models are well-understood and studied, the conceptual modelling errors are much more difficult to address, because a deep understanding of the model creation process is required.

From the mathematical perspective, enhancing our understanding of the modelling process implies a more formal description of mathematical models, the relations between them, and the coupling of models. Furthermore, this formal approach can be extended to develop tools for detecting conceptual modelling errors. Consequently, the goal of this book was to develop an abstract modelling framework based on category theory, supporting a understanding of the modelling process and providing a formal description of relations and couplings between models.

To better understand the difference between conceptual modelling and practical use of model, a two-phase modelling methodology, consisting of the conceptual modelling phase and the instance modelling phase was proposed in Chap. 1. Generally speaking, the conceptual modelling phase concerned with capturing all relevant aspects of the problem under consideration as mathematical formulations, while the instance modelling phase concerned with the solution procedure for these formulations, which can be done by means of analytic or numerical methods, or, in general, by means of simulations. Using this two-phase modelling methodology, it was clearly demonstrated that conceptual modelling and conceptual modelling errors significantly impact the quality of the modelling process outcomes. There-

D. Legatiuk, *Mathematical Modelling by Help of Category Theory*,
Mathematical Engineering, https://doi.org/10.1007/978-3-031-84951-0_6

fore, it is necessary to develop tools for supporting understanding of conceptual modelling.

Based on the above discussion, this book is intended for readers interested in gaining a deeper understanding of mathematical modelling, as well as those with an interest in formal languages and systems. It is also aimed at applied mathematicians and engineers involved in practical mathematical modelling. As typical in research, not all the results presented in this book are completed, leaving open possibilities for future research. It is hoped that these ideas will inspire readers to approach their (applied or theoretical) problems from a new perspective, leading to a deeper understanding and new insights.

Index

D. Legatiuk, *Mathematical Modelling by Help of Category Theory*,
Mathematical Engineering, https://doi.org/10.1007/978-3-031-84951-0

GPSR Compliance

The European Union's (EU) General Product Safety Regulation (GPSR) is a set of rules that requires consumer products to be safe and our obligations to ensure this.

If you have any concerns about our products, you can contact us on ProductSafety@springernature.com

In case Publisher is established outside the EU, the EU authorized representative is:

Springer Nature Customer Service Center GmbH
Europaplatz 3
69115 Heidelberg, Germany

Batch number: 10732044

Printed by Printforce, the Netherlands